Air and Water

Grenoble Sciences

The aim of Grenoble Sciences is twofold:
▶ to produce works corresponding to a clearly defined project, without the constraints of trends nor curriculum,
▶ to ensure the utmost scientific and pedagogic quality of the selected works: each project is selected by Grenoble Sciences with the help of anonymous referees. In order to optimize the work, the authors interact for a year (on average) with the members of a reading committee, whose names figure in the front pages of the work, which is then co-published with the most suitable publishing partner.

Contact:
E-mail: editions@univ-grenoble-alpes.fr
Website: https://www.uga-editions.com

Scientific Director of Grenoble Sciences:
Jean BORNAREL, Emeritus Professor at Grenoble Alpes University, France

Grenoble Sciences is part of **UGA Éditions**, a department of Grenoble Alpes University, and is supported by the **ministère de l'Enseignement supérieur, de la Recherche et de l'Innovation** and the **région Auvergne-Rhône-Alpes**.

This book is translated, revised and adapted from *L'air et l'eau – Alizés, cyclones, Gulf Stream, tsunamis et tant d'autres curiosités naturelles* by René Moreau, EDP Sciences, Grenoble Sciences Series, 2013, ISBN 978-2-7598-0828-1.

The reading committee of the French version included the following members:
▶ Jean BORNAREL, Emeritus Professor, Grenoble Alpes University,
▶ Jean-Pierre HULIN, Emeritus CNRS Senior Researcher, Fluids, Automatic and Thermal Systems Laboratory (FAST), Orsay,
▶ James LEQUEUX, Emeritus Astronomer, Paris Observatory,
▶ Jean LILENSTEN, CNRS Senior Researcher, Planetology and Astrophysics Institute of Grenoble (IPAG),
▶ Robert LUFT, Professor (retired), Nice Sophia Antipolis University,
▶ Jean-Yves MÉRINDOL, Professor, Strasbourg University,
▶ José TEIXEIRA, CNRS Senior Researcher, Léon Brillouin Laboratory, Paris.

Translation from original French version performed by Brett KRAABEL
Typesetting: Sylvie BORDAGE and Gwenn COGNARD
Graphic design and figures: Anne-Claire LECOMTE
Cover illustration: Alice GIRAUD, from: elements provided by the author, elements and creations realized by Alice GIRAUD

René Moreau

Air and Water

Trade Winds, Hurricanes, Gulf Stream,
Tsunamis and Other Striking Phenomena

 Springer

René Moreau
Laboratory of Science, Engineering,
  Materials and Processes (SIMAP)
Grenoble Institute of Technology
Grenoble
France

ISBN 978-3-319-65213-9   ISBN 978-3-319-65215-3   (eBook)
DOI 10.1007/978-3-319-65215-3

Library of Congress Control Number: 2017949117

Printed on acid-free paper

This Springer imprint is published by Springer Nature
The registered company is Springer International Publishing AG
The registered company address is: Gewerbestrasse 11, 6330 Cham, Switzerland

Reader's guide

The main goal of this book is to introduce the reader to the observable phenomena that occur in air and water. The paths leading to the comprehension of these phenomena are well marked and numerous examples accompany the reading along the way. More advanced notions that the author considers indispensable for understanding are presented in brief inserts set apart from the main text. These appear in the first few chapters and offer the keys to unlock any doors to comprehension that the reader may find locked. Footnotes also provide additional brief explanations, mostly concerning questions of etymology, bibliography, geography, or even history.

A heightened understanding of these two media, air and water, requires drawing from all scientific disciplines, and particularly that of fluid mechanics. However, this discipline is not covered in typical high school and university physics courses. Nevertheless, the author has made the wager of proposing in this book an introductory level of explanation that can be grasped without the benefit of advanced courses. Notions that deal with relatively subtle topics, such as the mechanisms behind hydrodynamic instabilities and turbulence, are therefore placed in the appendix, which constitutes a sort of common base for the eight chapters of the book, which themselves contain no equations or abstract concepts. Finally, because of the abundant vocabulary required to describe these phenomena, a fairly detailed glossary is available at the end of the book. In it, the reader will find the meaning of the majority of the scientific terms used in the book. Upon the first use of such a term in the main text, the epilogue, or the appendix, it appears in blue text so as to inform the reader that its meaning is to be found in the glossary.

Acknowledgments

To start with, I must thank the undergraduate students, engineering students, doctoral students, and colleagues who, all along my career, encouraged me with their questions and comments to find explanations as simple as possible for all the phenomena that we observe in fluid media such as air, water, and many others as well. By bringing me realization that the behavior of these fluids could be described with everyday words, it is they who allowed me to imagine this book.

My thanks also go to the members of the reading committee. Their very constructive suggestions led to a book that is significantly better than the initial manuscript.

I also must thank the people, photographers, companies, and organizations that graciously provided, or helped me obtain, numerous photographs and illustrations showing both spectacular natural phenomena and large engineering works. If the presentation of this book is found appealing, it is thanks to their contributions.

Finally, I am particularly indebted to the team at Grenoble Sciences who, under the direction of Jean Bornarel, put this book into its final form, which was no easy task considering the inserts and the numerous figures. Sylvie Bordage, Laura Capolo, Gwenn Cognard, Anne-Laure Passavant and Stéphanie Trine, as well as Anne-Claire Lecomte and Alice Giraud exhibited real professionalism and a lot of patience with me over the months before the book came out. I also wish to express my gratitude to Brett Kraabel for his efficient and friendly translation from the French text into this English version.

Table of contents

Prologue

At the beginning of all sciences is the astonishment that things are as they are.

(ARISTOTLE)

Air and water are two fluids essential for life. Encountered in the first lessons of childhood, they are so familiar that we all think we know them. Yet how difficult it remains to predict their behavior, with so many questions butting against the limits of our knowledge. Can we explain with precision the wind, thunderstorms, and hurricanes? Why does it rain here and not there? Will next summer send a heat wave, or will it be memorably pleasant? Will the sea be calm or raging? And where do these waves come from that endlessly break on our shores?

As part of their education, engineers and technicians in training have the opportunity to visit large hydraulic facilities, ports, or airports, allowing them the chance to train their eye on these fluid environments in permanent motion. As their understanding of the origins of these phenomena improves, so does their admiration for the prodigious show offered by this vigorous nature. Progressively, their interest grows and, in return, their comprehension of these phenomena cultivates within them a real desire to master the tools that allow them to be analyzed. Even visitors to these spectacular facilities from outside science and technology are blessed with this attitude. Their guides, accustomed to explaining the power of falling water without equations or abstract concepts, must lead them in an approach that requires looking, if not observation, which not only engages much more but also teaches much more than simply seeing.

Be it air or water, the voyage proposed to the reader begins with a presentation of an environment supposedly at rest, then continues by examining its incessant pulsations, difficult to predict. First focusing on the largest scale, that of the entire planet, the itinerary then leads to ever smaller structures, such as low-pressure zones in the atmosphere, clouds, rain, as well as tides and waves. We travel also along the immense network of waterways that drains and irrigates the continents, along the way admiring the magnificent works such as dams and asking how they impact the surrounding territories. The central idea behind this book consists thus of observing natural phenomena and proposing relatively simple explications destined to be understandable to a large public, concerned about their environment.

Most of the text is accessible to readers with no more than high-school science and who are at ease with quantities such as the temperature of a fluid or the pressure within such a medium. However, explaining certain phenomena requires notions more advanced than are encountered at the undergraduate level. To not perturb our trip, these notions have been isolated into inserts in the associated chapters. The noninitiated reader should pass over these without worry, knowing that the notions will be acquired through examples.

In the main text equations are thus abandoned, although some essential quantities are given to fix the orders of magnitude of the phenomena discussed and establish some useful benchmarks. As an example, consider the characteristic timescales of atmospheric movements versus those of the oceans. The former are at most on the order of weeks whereas the latter exceed the millennium: the large thermohaline circulation takes about 1600 years to circle the globe! Motivated by the interactions between the atmosphere and the oceans, we shall discuss meteorology and visit the theme of climatology, which is the subject of intense research in these times of global warming. The goal is to identify the scientific basis of these large challenges by limiting ourselves to discussing their well-established aspects, but without treating them in great detail.

In essence, I hope to give the reader feeling that I am telling a beautiful story and of transmitting a sense of my awe and wonder of it all. From this viewpoint, I hope I may be allowed to invoke Jules VERNE, who in my eyes is one of the masters of the attitude of awe and stupefaction when confronted with scientific knowledge. His *Extraordinary Voyages* so marked the child that I was, that I feel the need to cite the two volumes closest to the purpose of this book and from which I borrowed the opening quotes for chapters 2 and 5: *Five Weeks in a Balloon* and *Twenty Thousand Leagues under the Sea*. Admittedly, his vision was based more on dreams than on scientific comprehension. After having read this rational description of phenomena observed in air and in water, I hope that this book helps readers follow their own course, mixing dream with reality.

The author

Chapter 1

The atmosphere at rest

The air is full of the shudders of things that flee.

(Charles BAUDELAIRE, *Parisian Scenes*)

© Springer International Publishing AG 2017
R. Moreau, *Air and Water*,
DOI 10.1007/978-3-319-65215-3_1

P lanet Earth is shaped like a sphere slightly flattened at the poles. Its average radius is about 6370 km and the perimeter at the equator is nearly 40 000 km. The atmosphere, filled with this marvelous gaseous mixture that we call *air*, is its outer envelope where humanity lives along with numerous species of animals and plants. We ceaselessly breathe this air, move in it, look through it, and we accept the vagaries of weather. As familiar as it seems to us, as indispensable to life as it may be, do we really know this atmosphere? The knowledge gained through the centuries and refined with ever greater precision during the aerospace age gives us now a good knowledge of this medium and allows us to understand its main properties. Let us set off on its observation then, limiting ourselves in this chapter to a first approximation: that of an atmosphere at rest.

1. Structure of the atmosphere

Far above Earth's surface, the boundaries between atmospheric layers become blurred in two zones where the gas is so rarified that approximating it as empty space is often justified: the magnetosphere and the heterosphere (see insert I1.1). As high as it may be, the magnetosphere, through which Earth's magnetic-field lines loop, does have an impact on Earth because it shields us from the life-threatening cosmic rays from the Sun. Blown by the cosmic wind,[1] its thickness varies greatly from several thousand kilometers for the dayside magnetosphere to much greater distances on the nightside as a result of the wake created by Earth in the cosmic wind. As for the heterosphere, where the concentration of the diverse particles becomes extremely small (over 10^5 times less than at the surface of Earth), it forms the transition between outer space and Earth itself. We will not discuss the properties of these far-off regions[2] but rather limit our observations to the densest and nearest part of the atmosphere, which is now called the *homosphere* and which is several hundred kilometers thick. This part of the atmosphere is relatively well known because it is constantly traveled through by numerous objects, such as airplanes and weather balloons.

---

1   The solar wind is a flux of elementary particles emitted by the Sun, mostly consisting of protons and electrons. The energy of the heaviest particles, the protons, ranges from 100 eV, which corresponds to a speed of about 500 km s^{-1}, to 10 MeV, which corresponds to 5000 km s^{-1}. The electron volt (eV) is a unit of energy that is not part of the SI system but is convenient for the physics of elementary particles (1 eV is equivalent to 1.60217653×10^{-19} J, 1 keV is 10^3 eV, 1 MeV is 10^6 eV). The Aurora Borealis is produced by the entry into the atmosphere of the rare particles that manage to breach the magnetosphere. This phenomenon occurs only in the polar regions where Earth's magnetic field is almost perpendicular to the ground.

2   Those interested in the magnetosphere and the heterosphere may consult the book *Space Weather, Environment and Societies*, by Jean LILENSTEN and Jean BORNAREL, Springer (selected by Grenoble Sciences), 2006.

Were Earth scaled down to the size of an apple, the thickness of the homosphere would be reduced to a fraction of a millimeter: the skin of the apple! In this layer of essentially gaseous matter pinned by gravity to the surface of Earth, we can distinguish three sublayers whose properties are dominated by different physical mechanisms: the mesosphere, which is uppermost, constitutes the transition to the heterosphere; the troposphere, which is in contact with the surface; and the stratosphere at intermediate distances. In everyday language, the *lower atmosphere* refers either to the troposphere or to the entire homosphere. Figure 1.1 shows how temperature varies with altitude in these three layers and beyond. The interfaces between the layers, called the *stratopause* and the *tropopause*, are regions of transition between the dominant physical mechanisms of the corresponding layers. Without going into details that can be found in more specialized texts, we will more closely examine these physical mechanisms and their effects. Note first, however, that the troposphere alone contains over 80% of the total mass of the atmosphere, and that half of this mass falls within 5500 m from sea level.

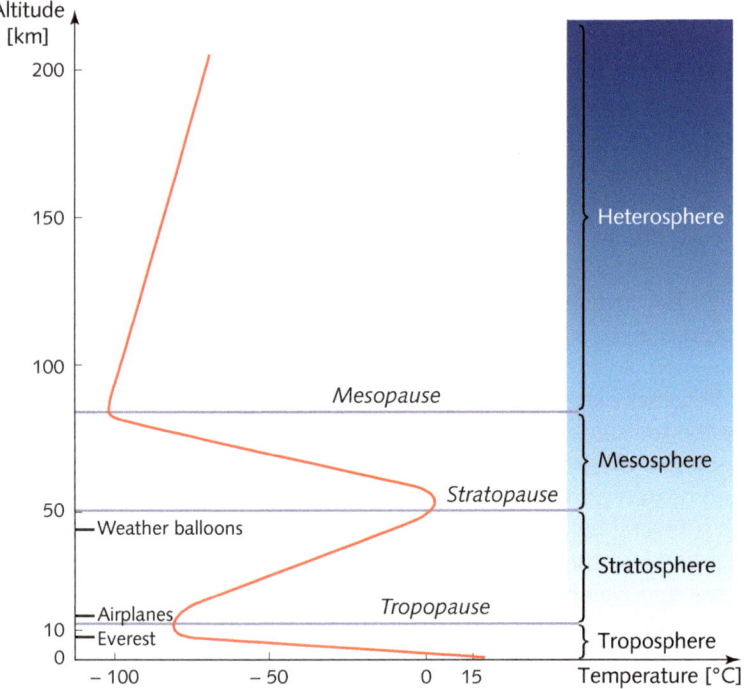

Figure 1.1 - Typical temperature distribution as a function of altitude across the concentric layers that make up the homosphere: the troposphere, stratosphere, and mesosphere. In the troposphere the drop in temperature with altitude is due to the fact that this layer is essentially heated by the ground via conduction and convection. The rise in temperature in the stratosphere comes from the absorption of ultraviolet radiation from the Sun, which leads to the formation of the ozone layer. The temperature drops again in the mesosphere, which is deprived of any significant heat source. [© NASA]

I1.1 - A bit of etymology

To physically describe the Earth requires a precise vocabulary whose words are built upon Latin or Greek roots of appropriate meaning. The principal roots are three Greek words: **sphaira** (σφαῖρα, ball), which become *sphere* in English and can designate regions not necessarily spherical but closed onto themselves, **genos** (γένος, origin), from which comes the noun *gene*, and the verb **skopein** (σκοπειν, to observe), which is much used in optics and in observation or measurement. Another important root is the Latin word **pausa**, which means *stop*. Complementing these are Greek prefixes, such as *homo* from **homos** (ὁμός, the same), *hetero* from **heteros** (ἕτερος, other), *macro* from **makros** (μακρός, large), *meso* from **mesos** (μέσος, intermediate), *micro* from **mikros** (μικρός, small), and *tropo* from **trepein** (τρέπειν, change). Two more Greek prefixes of importance are *magneto*, which comes from **magnes** (μάγνης, a stone from Magnesia; a magnetic mineral), and *strato* from **stratos** (στρατός, layer). A large part of the geophysics vocabulary is derived from this etymology, in particular the nouns that designate terrestrial atmospheric zones, such as magnetosphere, heterosphere, homosphere, mesosphere, stratosphere, troposphere, stratopause, and tropopause. Three adjectives that identify the scale of observations or that describe phenomena have become quite common beyond the framework of physics: macroscopic, mesoscopic, and microscopic.

The troposphere is characterized by a regular and monotonic reduction in temperature and pressure with altitude. At sea level, the average temperature is about 15 °C, the average pressure is about 1013 hectopascals (hPa) or millibars, and the density is about 1.2 kg m^{-3} or about 800 times less than liquid water. The temperature[3] decreases almost linearly with altitude, losing on average about 6.5 °C per kilometer until it reaches −56 °C at the tropopause. This reduction in temperature results because the troposphere is essentially heated by conduction and convection from the ground and oceans, which obliges the air at all altitudes to be colder than the air below yet warmer than the air above. The ground and oceans are heated directly by solar radiation that penetrates the atmosphere, which is essentially transparent. Nevertheless, as we shall see later in this chapter (section 1.4: *Heat exchange via the atmosphere*), the troposphere also intercepts a fraction of the infrared radiation emitted by Earth, leaving the balance to escape into space.

---

3   In this book, temperature is expressed in degrees CELSIUS (°C). To convert to absolute temperature, which is counted in kelvins (K) in honor of the 19th century Scottish physicist William THOMSON, 1st Baron KELVIN (1824−1907), you simply add 273.15 to the temperature given in degrees CELSIUS. In relative terms, the variation in absolute temperature across the troposphere, which goes from 288 K on the surface to 217 K at the tropopause, is a fairly modest 25% compared with the variation in pressure, which is around 100%. This is why a common first approximation when considering only the lower layers of the troposphere is to assume that the density ρ can be deduced from the pressure p with the temperature taken as constant. This is done by using the BOYLE-MARIOTTE's law $p/\rho = const.$ For better precision, the equation of state must be considered, as we shall see later.

The linear decrease in temperature within the troposphere reflects the overriding importance of conduction and convection in this relatively dense medium. The average rate of decrease of 6.5 °C km^{-1}, which reflects the apparent thermal conductivity of air in this zone, can vary significantly from 5 °C km^{-1} in air saturated with water vapor to 9 °C km^{-1} in dry air.

Unlike temperature, pressure decreases exponentially with altitude. This property results from the required equilibrium imposed by gravity on the entire troposphere, which can be expressed in a simple way. All volumes, such as the slab $ABCD$ shown in figure 1.2, are in equilibrium when acted on by two opposing forces of equal magnitude. The downward weight of the air in the slab $ABCD$ must therefore be exactly countered by the force due the difference dp in pressure between the planes AB and CD.[4] The pressure at AB must be larger than the pressure at CD because pressure decreases with altitude, so dp clearly acts upward on the slab. In the troposphere, gravity and the absolute temperature vary slowly enough that, at least to a first approximation, we can consider them to be constant over small distances. In addition, because the equation of state of atmospheric gas is close to that of an ideal gas,[5] its density varies approximately proportionally to the pressure. As a result, the local variation in pressure is essentially proportional to the pressure itself, which is the trademark of exponential functions.

This implies that, at a given altitude in the troposphere, the fall in pressure is not the same across all layers of equal thickness. The decrease is about 110 hPa km^{-1} at sea level,[6] about 70 hPa km^{-1} at the altitude of Mont Blanc (4810 m), and some 40 hPa km^{-1} at the altitude of transatlantic flights (around 10 000 m). Note that, at this altitude, the air is sufficiently dense to provide lift to the wings of airplanes but, beyond 15 000 m, the air is five times less dense than at sea level. Thus, all else being equal, generating lift for an airplane above 10 000 m is more difficult due to the thin

---

4   This example shows the importance, and also the limitations, of the very general and powerful approach used to explain physical phenomena; namely, to simplify as much as possible without losing the essential ingredients and falsifying the results. In reality, an astronomical number of rapidly and randomly moving molecules inhabit the slab, so the system is not strictly in equilibrium. Nevertheless, by modeling this layer as a medium at rest distinct from the surrounding medium, it soon reveals its macroscopic behavior to be characterized by an exponential decrease in pressure. This viewpoint and its limits are discussed in the book *Thinking in Physics. The pleasure of reasoning and understanding*, by Laurence VIENNOT, Springer (selected by Grenoble Sciences), 2014.

5   Using p to denote pressure, ρ for the density, and T for the absolute temperature, the equation of state of an ideal gas is $p/\rho = RT/M$, where R, the universal gas constant, weighs in at 8.314472 joules per mole and per kelvin ($R = 8.314472$ J mol^{-1} K^{-1}) and M denotes the molecular mass of the gas (i.e., the average molecular mass for a mixture of gas such as air).

6   The international unit for pressure is the pascal (1 Pa = 1 N m^{-2}), named in honor of Blaise PASCAL (1623–1662), the French scholar and moralist of the 17th century who was among the first to extract the concept of pressure. A pascal is about 10^5 times less than atmospheric pressure at sea level, which is about 10^5 Pa. As a result of the law of hydrostatics $p + \rho gz = const.$ (ρ is the density, g is gravity, and z is altitude), this pressure applied on the ground by the entire height of the atmosphere is equivalent to the pressure exerted by a 10-m-high column of water or a 76-cm-high column of mercury.

air; however, the airplane will also experience less drag and so can fly much faster[7] at a given rate of energy consumption. This explains why vehicles conceived to fly at such altitudes have done away with the large wings and retain only small stabilizers—these are rockets, which propel and orient themselves with their engines.

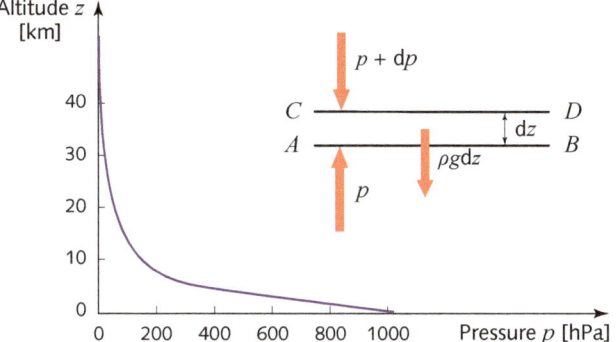

Figure 1.2 - Exponential decay of pressure in the troposphere, which results from the equilibrium between the forces of pressure on AB and CD (the net force on a unit area due to pressure is dp) and the weight on the volume $ABCD$ of air with unit length and thickness dz (ρ is the density, g is gravity, and z is altitude).

To get a more concrete feeling of the effects of this rapid decrease in pressure across the troposphere, imagine a climber that starts a trek at sea level. When he arrives at the summit of Mont Blanc, the pressure and density have decreased by almost half. Before each effort, our climber must breathe two times faster than at sea level to obtain the oxygen required for his body, and his heart rate must increase from 60 beats per minute to 120. This is possible for numerous people in good health and explains the high traffic at the summit of this iconic peak in the Alps. The high summits of the Himalayas culminate around 8000 m, where the pressure and the density have been divided by three. The climber that attains these altitudes must therefore breathe three times faster than at sea level and increase his heart rate to 180 beats per minute before even embarking on the climb, which makes this mission almost impossible. Only highly trained climbers that have managed to reduce their heart rate to less than 50 beats per minute at sea level can make such climbs without carrying extra oxygen.

Let us fill in some more this image of the troposphere. It's average thickness is of the order of 10 km. In addition to the gravity that pins it to the ground and the countering forces of pressure, it is also subjected to centrifugal forces[8] because it

---

7   However, with its triangular-shaped wing, which gives it the advantage of a significant lift per unit area of wing, the Concorde flew at supersonic speeds above the oceans at an altitude of 20000 m. Modern jet fighters, such as the F-22 Raptor or the Rafale, can fly at nearly 25000 m.

8   In a rotating solid body such as Earth, the centrifugal force is proportional to the square of the angular speed and the distance from the axis of rotation; thus, the force is greatest at the equator

spins with Earth about its axis and to an upward momentum at the equator and in other zones due to an ascending fluid flow within the atmosphere. These forces are much greater at the equator than in the polar regions, where they disappear. Considering these effects, which vary with latitude, the real thickness of the troposphere goes from 7 to 8 km near the poles to 15 km at the equator.

This averaged description is perturbed and even significantly disrupted by meteorological phenomena that we will discuss in the next two chapters. Nevertheless, it is already possible to get a relatively good estimate of the magnitude of these disturbances. In the storms that hit Europe, winds frequently reach speeds of the order of 150 km h^{-1}. The BERNOULLI's[9] equation allows us to estimate the low pressure that is associated with these winds to be about 1040 Pa. This means that, at the center of such storms, the atmospheric pressure is reduced by 1% with respect to the pressure at the periphery of the storm. This relatively small difference is what justifies the assumption used in this chapter of an atmosphere at rest and the use of average values for pressure, density, temperature, and concentrations of chemical species instead of local, instantaneous values that fluctuate. Inversely, because pressure perturbations of the order of 1% already suffice to provoke relatively strong storms, we can already prepare ourselves to understand why meteorological fluctuations can develop so easily in such a light, airy medium.

2. Composition of the atmosphere

What is the composition of the air we breathe and whose pollution we fear? To a first approximation, dry air is a mixture of four gases: nitrogen, oxygen, argon, and carbon dioxide. They are present at sea level at concentrations of 78.1% for nitrogen (N_2), 20.9% for oxygen (O_2), 0.93% for argon (Ar), and 0.034% for carbon dioxide (CO_2). Air also contains what are called *minor traces* of other gases whose concentrations are much less (less than 1 part in 100 000): neon, helium, methane, krypton, hydrogen, nitrous monoxide, xenon, ozone, and nitrogen dioxide, as well as traces of carbon monoxide (CO) and ammonia (NH_3).

Of course, except very far from water surfaces and the consequent evaporation, air is rarely dry. Its water-vapor content, a parameter to which terrestrial life is very sensitive, varies from zero to a maximum, called the *saturation vapor pressure*,

---

and goes to zero at the poles. More information on the centrifugal force and on the CORIOLIS force, both of which are important in the mechanics of rotating bodies, is available in insert I2.2.

9   Denoting pressure by p, density of a fluid (air in this case) by ρ, gravity by g, altitude by z, and the speed of fluid flow by V, BERNOULLI's equation $p + \rho gz + \rho V^2/2 = const.$ allows us to compare the pressure of moving air with that of air at rest. At sea level ($z = 0$), with $\rho = 1.2$ kg m^{-3} and $V = 150$ km h$^{-1} \approx 42$ m s^{-1}, the difference in pressure between the exterior of a low-pressure zone (where $V = 0$) and the center is of the order of 1040 Pa. This relationship is named in honor of Daniel BERNOULLI (1700–1782), a Swiss doctor and physicist of the 18th century. Note that, for zero speed, the BERNOULLI's equation reduces to the law of hydrostatics $p + \rho gz = const.$

above which water condenses into liquid droplets. To characterize the humidity in air, a common approach is not to use directly the concentration or the partial pressure of water vapor in the air but instead to use the relative humidity, which is the ratio of the partial pressure of water vapor to the saturation vapor pressure and varies from 0% to 100%. This metric, however, can be somewhat misleading, because 100% of a very low saturation vapor pressure can lead you to believe that the moisture content of the air is high, whereas the air would actually be quite dry. In fact, the notion of humidity in air is itself difficult to reconcile with our senses. In winter, when the air is cold and full of clouds or fog, the gaseous fraction of the air is almost dry and the moisture is concentrated in water droplets or in snowflakes. Conversely, in the summer on a warm sunny day, the gaseous fraction of air contains a lot of water vapor and is quite humid; this is in fact the dampness we feel during hot, humid weather and in tropical climates.

The saturation vapor pressure is an important parameter for understanding cloud formation. It depends essentially on temperature and can in practice be deduced from the empirical formula developed by RANKINE.[10] To get a feeling for it, the saturation vapor pressure is listed here in hPa (or millibars) for several temperatures: 1 at −20 °C, 3 at −10 °C, 6 at 0 °C, 12 at 10 °C, 23 at 20 °C, 42 at 30 °C, and 74 at 40 °C. These numbers show that the saturation vapor pressure is maximal in hot air, since it can reach 74 hPa, which means that the concentration of water vapor in the air is about 7%. At the opposite extreme, in cold air, the saturation vapor density approaches zero at 0.6% of the standard atmospheric pressure at 0 °C.

Based on these first elements, certain well-known effects can be explained rather simply. During summer nights, on the shores of the Mediterranean, drying laundry is difficult despite a rather high temperature because the air is already saturated with moisture. In contrast, the moisture in the air can be deposited as dew on cool surfaces. During the day, however, the sunshine increases the air temperature, which is accompanied by an increase in the saturation vapor pressure. This leads to a decrease in the relative humidity, increases the capacity for the air to hold water vapor, and facilitates rapid evaporation of moisture in the laundry. Add to this that the sunshine directly absorbed by the laundry increases its temperature with respect to the ambient air, causing an even faster evaporation. This last remark explains why laundry hung in the Sun dries faster than laundry hung in the shade.

In the example of the wet laundry, the diurnal variations in temperature and sunshine are sufficient to explain the different times required to evaporate the water from the laundry. Other situations exist wherein the dryness of the air can be explained by the absence of a source of moisture. Thus, in continental regions such as the plains of central Europe, which are devoid both of large lakes and of

---

10 RANKIN's formula, $\ln p_{sat} = 13.7 - 5120/T$, where p_{sat} is in atmospheres (1 atm = 1013 hPa or millibars) and T is in kelvin, allows the saturation vapor pressure of water in air to be estimated. This equation is named in honor of William John MACQUORNE RANKINE (1820−1872), a Scottish physicist who was a pioneer in thermodynamics.

any significant vegetation cover, the dryness is due to the fact that the intense diurnal evaporation is never compensated by any significant inputs of moisture. As another example, consider high mountain ski resorts. Here, the saturation vapor pressure can become very low because of the low temperatures, which explains the very dry air in these places and the subsequent damage to the epidermis and to wood furniture.

At higher altitude is the stratosphere, which is three to four times thicker than the troposphere but contains only 20% of the total mass of the atmosphere. It can be considered as a rarified gas for which the approximations of classical aerodynamics must be revisited. The first of these approximations to become questionable is related to the hypothesis that the mean free path of molecules[11] is much shorter than the length scales of airflow, such as the size of an eddy or of an airplane in motion. This comparison of the characteristic length scale of molecular motion, which is a microscopic phenomenon, with the characteristic length scale of airflow, which is a macroscopic phenomenon, involves a very important and intuitive concept in physics; namely, the order of magnitude. This concept allows us to classify phenomena into different categories depending on their typical scales, be it a time scale, a length scale, a scale of speed, etc. (see insert I1.2). In addition, by invoking the mean free path of molecules in air, we are basing our discussion on the kinetic theory of gas, which allows parameters describing the microscopic world of molecules to be related to those that describe macroscopic world (see insert I1.3).

The stratosphere is thus a region where phenomena that require a rather dense medium are much less significant than in the troposphere. The pressure, which is already diminished to about 250 hPa in the troposphere, diminishes further to as low as 2 hPa in the stratosphere. Compared with the troposphere, the air in the stratosphere has almost no heavy constituents, such as liquid water. The very rare and extremely tenuous stratospheric clouds that we observe in the polar regions contain water and nitric acid.

Among the main characteristics of the stratosphere is the presence of an ozone layer, which is associated with a remarkable increase in temperature from $-56\,°C$ to about $0\,°C$. The ozone (O_3) layer is present between 20 and 30 km in altitude and is the result of absorption by oxygen molecules (O_2) of ultraviolet solar radiation, which splits oxygen molecules into two separate oxygen atoms. Ozone molecules are the final product of complex chemical reactions in which an oxygen atom combines with O_2 to form O_3. Although the lifetime of ozone in the stratosphere is longer than in the troposphere, it is not very stable in the stratosphere and decays naturally, producing O_2 again. The stratosphere is estimated to contain 90% of the atmospheric ozone, the rest being found in the lower layers of the troposphere

---

11 The mean free path of a molecule is the mean distance it covers between successive collisions with other molecules. This concept is discussed in more detail in insert I1.3, which gives a simple explanation of how the parameters of a molecule's movement can be connected to macroscopic quantities.

near large urban centers. The ozone-oxygen cycle is closely followed by meteo-rologists, on the one hand because, by absorbing a large part of the ultraviolet solar radiation, ozone plays an essential role in maintaining life on Earth and, on the other hand, because its chemistry depends on the concentration of pollutants in the atmosphere, such as nitrous oxides (NO_x) and chlorofluorocarbons (CFCs). The variations in the ozone layer, which seem to have a distinct seasonal character, have come to be called the *hole* in the ozone layer.

Overall, the dissociation and recombination reactions of O_2 and O_3 molecules are exothermic and warm the air in the high altitudes shown in figure 1.1. This increase in temperature in such a dilute gas, which betrays an increased agitation of the molecules, does not imply that macroscopic objects in the stratosphere, such as weather balloons, are as hot as the surrounding air. Because the impact of mol-ecules with such objects is relatively rare, they receive and stock little heat, whereas they lose heat by emitting infrared radiation. This increase in temperature of the stratosphere, with its concomitant decrease in density, renders this layer more stable than the troposphere, which is heated from below and subject to RAYLEIGH-BÉNARD convection (see appendix). This explains the calm and stratified structure of the stratosphere seen in figure 1.3.

Figure 1.3 - Photograph of the atmosphere at sunrise obtained by a NASA crew from the International Space Station in June, 2011. Given the scale, the blue layers are the stratosphere, and the more intense strip of blue is the ozone layer. The reddish layers above the dark ground are the troposphere. [© NASA]

The mesosphere, inaccessible to weather balloons and *a fortiori* to planes but tra-versed and travelled through by numerous artificial satellites, is less well known than the troposphere and the stratosphere. We can consider it as the transition to the heterosphere and to outer space. The heat given off by the formation of ozone, which is characteristic of the stratosphere, become negligible here, leading again to a strong decrease in temperature, as seen in figure 1.1. Conversely, variations in solar activity, whose effects are strongly moderated in the stratosphere and troposphere, are reflected in the mesosphere by significant phenomena, notably involving the concentration of diverse species and the dissociation of their molecules into free atoms.

In reality, the properties of the atmosphere are subject to significant intermittent variations, especially in the troposphere, which is influenced by the seasons, the diurnal cycle, and hydrodynamic instabilities. These variations led the International Civil Aviation Organization (ICAO) to define a standard atmosphere. All magnitudes measured under given conditions, themselves measured, are transformed into their equivalent in a standard atmosphere. Thus, all aeronautical records are defined and recorded for this standard atmosphere. The definition can be summed up as follows: altitude is measured from the average sea level, which constitutes a reference surface where the pressure is fixed at 1013.25 hPa, the temperature at 15 °C, and the density at 1.225 kg m^{-3}. The decrease of pressure with altitude is defined as shown in table 1 and the associated temperature follows linearly at 6.5 °C km^{-1}.

I1.2 - The concept of order of magnitude

Although the science of physics normally calls for the use of precise data, we sometimes consciously use approximate values, called the *order of magnitude*. The former is given by numbers containing several significant digits, which means we are sure of their precision (at least up to the least significant digit, which may be rounded). When making measurements, the number of significant digits grows with the progression in data-acquisition techniques. The order of magnitude plays a different yet complimentary role; it is no substitute for precise values. Instead, it serves to categorize phenomena; for example, to separate slow phenomena from fast, or long length scales from short.

Take the example of speed. The speed of a person walking is about 3 km h^{-1}, that of a cyclist is some 30 km h^{-1}, and that of a high-speed TGV (*Train à Grande Vitesse*) train is 300 km h^{-1}. As they travel, the walker has time to look at the flowers and trees, the cyclist has just enough time to read the street signs, and the TGV passenger gets only a global view of the countryside and is completely oblivious to the small-scale details. In addition, the walker can cover around 10 km in a day, the cyclist a hundred, and the TGV passenger can go from Paris to London in 3 hours, allowing him several hours at his destination while still being able to return home for dinner. Without making any precise measurements, we can identify in these examples a slow speed, a medium speed, and a fast speed.

When using the order of magnitude in physics, we go a step further and delete all significant digits, keeping only the relevant power of ten: 1 km h^{-1} for the walker, 10 km h^{-1} for the cyclist, and 100 km h^{-1} for the train. This retains the distinguishing feature of each speed. In addition, we put all walkers into the 1 km h^{-1} category, all cyclists into the 10 km h^{-1} category, and all trains in the 100 km h^{-1} category. This approach, which allows us to sort speeds by category, also has the advantage of ridding the phenomenon in question from distracting details, such as the significant digits. However, we cannot use the order of magnitude to distinguish between phenomena within the same order (e.g., one person who walks faster than another). This requires precise data.

Throughout this book, we shall find the need to discuss and compare orders of magnitude; for example, to distinguish between rapid phenomena, such as the atmosphere that evolves on the timescale of days, from slower phenomena, such as ocean currents whose timescales exceed a millennium. And let us not neglect the fact that many other disciplines, such as the social sciences or economics, also use orders of magnitude, even if the approach is not tacitly codified as in physics.

Table 1 - Pressure as a function of altitude in the standard atmosphere defined by the International Civil Aviation Organization (ICAO).

Altitude [km]	Pressure [hPa]	Temperature [°C]
0	1013	15
1	900	8.5
2	794	2
3	700	− 4.5
4	617	− 11
5	541	− 17.5
6	471	− 24
7	411	− 30.5
8	357	− 37
9	307	− 43.5
10	265	− 50
11	227	− 56.5
12	194	− 56.5
13	165	− 56.5
14	141	− 56.5
15	119	− 56.5
20	55	− 46
30	11	− 38
40	3	− 5
50	0.9	+ 1

I1.3 – Glimpse into the kinetic theory of gas applied to the atmosphere

Air at rest is composed of molecules in permanent motion, exchanging their momentum through their frequent collisions. On the microscopic scale, the parameters describing this motion differ vastly (by several orders of magnitude, as we shall see) from any eventual macroscopic motion. This justifies the postulate that, even when considering macroscopic motions, air is almost in equilibrium. In addition, the kinetic theory of gasses allows us to understand and to model macroscopic quantities, such as pressure, temperature, concentrations, and all the physical properties of this gas. Although this book is not the place to explain the details of the kinetic theory of gasses, which is covered in many specialized texts, we can, if we accept certain approximations, understand in terms of order of magnitude the relationship between the parameters of the molecular world and the macroscopic quantities.

Let us assume that all molecules in air are identical spheres with radius σ, mass m, and speed c, and that the gas contains n molecules per unit volume. To start with, we can express the density of this gas as $\rho = nm$. Also easily done is to determine the mean free path of these molecules (let d be the average distance between successive collisions) by dividing the distance traveled by one molecule per unit time by the number of molecules in the cylinder swept out by this molecule. The result is $d = 1/n\pi\sigma^2$. We can then define the pressure as the cumulated impulses of all the molecules that collide with the plane wall of the (real or imaginary) container per unit area and per unit time. This calculation is not trivial, because it implies summing the contributions of molecules that come from all directions. Nevertheless, it leads to the simple result $p = nmc^2/3$. This result seems reasonable because the pressure should be proportional to n, m and c^2, so the only contribution of the calculation is the factor of 1/3.

Since density and pressure are linked to the molecular parameters, obtaining the temperature requires nothing else than to identify this pressure as the pressure that appears in the ideal gas law $p/\rho = RT/M$ (where R is the universal gas constant and M is the average molecular weight of the mixture). This leads to the expression $T = Mc^2/3R$, better known in the form $k_B T = mc^2/3$, where we have introduced BOLTZMANN's constant $k_B = 1.38041 \times 10^{-23}$ joules (related to R and to AVOGADRO's number N_A by $k_B = R/N_A$, with $N_A = 6.0248 \times 10^{23}$ mol^{-1}). Finally, the dynamic viscosity μ of the gas, which expresses the net result of the momentum exchanged by the molecules across a given plane, is $\mu = nmcd/3$.

In addition to these relationships, note that pressure, temperature, and the transport properties of a gas are macroscopic manifestations of the extremely fast microscopic agitation of the molecules. One point of interest for these relationships is that knowledge of the macroscopic magnitudes leads to a relatively good approximation of the parameters describing the molecular world. For example, with $p = 10^5$ Pa, $T = 300$ K, $\rho = 1.3$ kg m^{-3}, $\mu = 1.8 \times 10^{-5}$ kg m^{-1} s^{-1}, we find $n = 2.4 \times 10^{25}$ m^{-3}, $m = 5.4 \times 10^{-26}$ kg, $c = 480$ m s^{-1}, $d = 4.2 \times 10^{-7}$ m, and $\sigma = 1.8 \times 10^{-10}$ m.

3. Propagation of waves in the atmosphere

3.1. Sound

This air that we are beginning to know possesses two more very familiar properties: sound and light can propagate through it. Let us start with sound, which leads us to examine the nature of acoustic waves. We have just seen that, in a gas such as air that is said to be *at rest*, the constant and rapid agitation of molecules translates on macroscopic scales into properties that we can directly detect with our senses, such as temperature and pressure. These properties are measurable and are called *state variables* because they characterize the overall state of the medium, which is at rest macroscopically but highly agitated on the microscopic scale (see insert I1.3). The absolute zero of temperature (0 K), which corresponds to the unrealistic case of an immobile gas completely devoid of internal energy, has no practical meaning because the gaseous state exists only above a certain nonzero temperature characteristic of the equilibrium between evaporation and condensation. By analogous reasoning, the concept of zero pressure (0 Pa) also loses all significance because it would correspond to a gas that asymptotically approaches infinite dilution; in other words, a total vacuum.

In the gaseous state (i.e., at nonzero temperature and pressure), a molecule that is shaken by, say, the vibration of a guitar string, immediately transfers this perturbation to the first molecule it meets which, in turn, transmits it to its neighbors, and so forth. In this way, acoustic waves propagate and can attain microphones or a listener's ear. From this initial idea of the physical nature of acoustic waves we can already deduce an important consequence: the celerity of sound waves must give a relatively good order-of-magnitude approximation of the typical speed of molecules as they move randomly about.

Under normal conditions, the celerity of sound in air is around 340 m s^{-1}, which differs little in terms of order of magnitude from the speed of 480 m s^{-1} for molecules in a gas deduced in insert I1.3 from the simplified molecular model. That this celerity is slightly less than the speed of the molecules themselves is reasonable because we cannot expect the random movements of molecules to transmit sound waves with 100% efficiency. In contrast, sound travels faster in denser media, where information can pass much more rapidly from one particle to the next. For example, in liquids such as water, the celerity of sound reaches 1500 m s^{-1}. In solids, which are both dense and well organized on atomic or molecular scales, sound can travel as fast as 4000 to 5000 m s^{-1}. At 340 m s^{-1}, the celerity of sound in air explains why spectators seated at a baseball game some 170 m from the hitter hear the crack of his bat about a half second after they see the impact of the bat on the ball. And over short distances, such as the dimensions of a concert hall, sound waves can transmit very subtle information, much to the delight of music lovers.

But sound propagated in this way loses its intensity fairly rapidly, both because its energy is distributed over an ever larger domain as the wavefront moves away from its source and because this initial energy is progressively dissipated into heat by the viscous friction within air. In the expanse of the atmosphere over a landscape without features, we can associate a well-defined range to sound from a given source, which of course explains why we must raise our voice more and more to be heard by a person moving away from us. Everyone also knows that walls, be they natural such as rocky cliffs or artificial such as the enclosure of a theater from antiquity, can significantly affect the signal received. These are complex phenomena whose analysis allows us to explain echoes and leads to important applications such as the design of concert halls with excellent acoustics or the implementation of systems to protect against noise pollution. In addition, the hearing of humans is sensitive to sound between the frequencies of 15 Hz[12] (the limit between infrasound and sound) and 20000 Hz (the limit between sound and ultrasound). These limits vary strongly between people and between animal species.

3.2. Light

The propagation of light through air is another very important phenomenon. Recall first the physical nature of this radiation. Light can be described either as an electromagnetic wave that propagates at 300000 km s^{-1} through a vacuum or as a flux of quantum particles called *photons*. Our eyes can detect light with a wavelength between 0.38 and 0.78 µm (1 µm, or *micron*, is 1/1000 of a millimeter), which corresponds to the visible spectrum[13] that goes from violet to red, respectively. Ultraviolet radiation (UV; wavelengths shorter than 0.38 µm) is strongly absorbed in the first few atomic layers of most materials, where the energy is converted into heat. This explains why, after prolonged exposure, UV radiation can burn the skin. As for infrared radiation (IR; wavelengths longer than 0.78 µm), it is known for the heat it deposits in the air through which it travels, which is less transparent to IR than to visible radiation. That IR can be directed to specific regions allows it to be used to heat a volume of air that is relatively static, such as the entrance of a commercial center or the terrace of a restaurant during a cool evening. A classic property of visible light is that each wavelength corresponds to a particular color, which is shown simply by passing a white light through a prism. For example, 0.5 µm light corresponds to the color green, which is appeasing to many and which lies in the middle of the visible spectrum.

---

12 One hertz (Hz) is the frequency corresponding to one complete oscillation (one period) per second. This unit was named in honor of Heinrich Rudolf HERTZ (1857–1894), a German physicist who specialized in electromagnetism.

13 The concept of a spectrum, or a spectral distribution, is common to all sorts of phenomena whose energy is distributed over a given range of wavelengths. A spectrum is the curve showing the contribution to the total energy from each wavelength. We often speak of spectral lines for phenomena whose energy is localized around one or several well-defined wavelengths and of continuous spectra for phenomena whose energy is distributed across relatively large wavelength ranges. The word *spectrum* comes from the Latin word *spectrum*, which means the simulacrum of an object.

Let us come back now to the baseball fan that was seated 170 m from home plate when he heard the crack of the bat hitting the ball about a half second after he saw the bat hitting the ball. The image of the hit, carried by light, took about 0.5×10^{-6} s to arrive on his retina. In contrast to the sound, we can consider that the light signal arrived essentially instantaneously. In addition to this enormous difference in speed, another major difference between light and sound is that light propagates through a vacuum whereas sound requires a medium through which to propagate: we have already seen this in air where sound is carried by successive collisions between molecules.

The most important source of natural light in the atmosphere is the Sun. As shown in figure 1.4, only a small part of the large spectrum of radiation emitted by our star is visible. The entire solar spectrum stretches from a wavelength of 10^{-1} μm to wavelengths greater than 10^4 μm. Altogether, the visible part of the solar spectrum is white, as determined by the elevated temperature of the Sun (around 5700 K), and is so intense as to be dangerous to our retinas. Fortunately, it arrives to us only after passing through several intermediary filters. The first of these filters is the atmosphere, often referred to as the *sky*. The color of the sky, which is due its illumination by the Sun, varies from pinkish at dawn to blue at midday to reddish before sunset, without including the various shades of gray due to clouds. The color also varies from one position in the sky to another: at midday, the sky directly overhead is a much deeper blue than at the horizon. Were it not for the atmosphere, we would live under a permanent night-time sky: black, sprinkled with stars that emit light and planets and moons that reflect some the light that strikes them. In spaceships orbiting above Earth's atmosphere, astronauts see this type of sky when they look toward the zenith; black because it absorbs almost all of the light it receives without reflecting any of it (see fig. 1.3).

To explain the colors of the sky, we must analyze the interactions between the incident white light and the molecules or particles in the terrestrial atmosphere, which intercept the photons, eventually withdraw some of their energy, and reemit them in all directions. Every type of molecule has specific properties of scattering and absorption, so the light reemitted by a volume of gas betrays the composition of the gas. Thus, the terrestrial atmosphere is blue because it is composed mostly of nitrogen and oxygen molecules, whereas the atmosphere of Jupiter is pinkish because it contains a large concentration of hydrogen, with an outer layer of ammonia. Explaining the interactions between the molecules in the atmosphere and light requires theories[14] based on quantum physics, which is beyond the scope of this

---

14 The English physicist John William Strutt RAYLEIGH (1842–1914) analyzed the scattering of light by atoms or molecules, and the German physicist Gustav MIE (1869–1957) analyzed light scattering by spherical particles such as water droplets. RAYLEIGH's theory shows that the efficiency with which light is scattered by particles much smaller than the wavelength λ of the incident light depends very strongly on the wavelength of the light (proportional to $1/\lambda^4$). Thus, blue light, having a shorter wavelength, is more efficiently scattered by nitrogen and oxygen molecules than is red light, which has a longer wavelength. In contrast, the scattering efficiency of water molecules is

book. Instead, we propose an explanation based on more elementary notions of the spectral distribution of light and on the phenomena of dispersion, absorption, and refraction of waves.

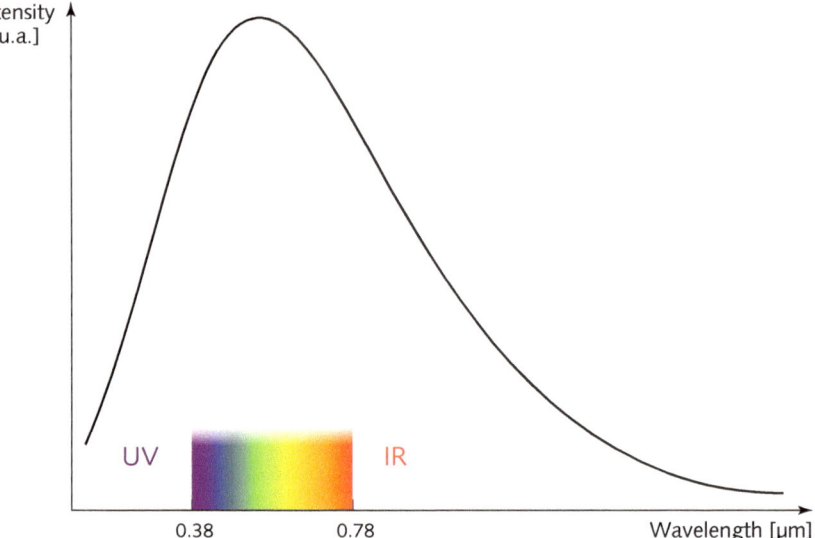

Figure 1.4 - Typical spectrum of radiation emitted from the surface of a body, such as the Sun, at a given temperature T. Between the ultraviolet and infrared parts of the spectrum is the visible band, ranging from violet to red. The black curve, expressed by PLANCK's law $b = A/\lambda^5(e^{c/\lambda T} - 1)$, gives the spectral brilliance b of a black body as a function of wavelength λ and for a given temperature T. The constants A and c are $A = 3.3745 \times 10^{-16}$ J m^2 s^{-1}, $c = 0.0144$ m K. The temperature T at the surface of the Sun is about 5700 K.

The color of the noon sky is due to the composition of the terrestrial atmosphere, which is rich in the diatomic molecules N_2 and O_2. These molecules more strongly scatter short-wavelength light (< 0.45 µm), which is close to the ultraviolet, and thus give off light that is more bluish than the incident light. In addition, the blue of the sky varies as a function of the fluctuating composition of its air. The color becomes more indigo for air that is dry and dust free, such as in the high mountains, more pale as the humidity increases, and yellowish or grayish for air containing a significant concentration of aerosols. From a simplified viewpoint, we can consider

---

roughly the same for all visible wavelengths. The presence of water vapor in the atmosphere thus washes down the deep-blue color of the sky, which is particularly noticeable in the high mountains where the air is dry rather than in the valleys where the air is more humid. MIE's theory explains that the scattering efficiency of spherical particles roughly the size of the wavelength of the incident light is independent of wavelength and can be accompanied by a strong absorption. Thus, the small water droplets in clouds strongly absorb the energy of the incident light, regardless of the particular wavelength, which explains why clouds reemit a white light of much weaker intensity than the incident light. Clouds carrying larger water droplets absorb a larger fraction of the incident energy, reemitting a gray light and becoming darker and darker as their water content increases.

that this shift toward pale blue in the light reemitted by the humid atmosphere is due to the fact that the efficiency with which water molecules scatter light is about the same across the entire visible spectrum, so these molecules contribute white light to the color of the sky. In contrast, the efficiency with which nitrogen and oxygen molecules scatter short wavelengths (blues) is higher than for the longer wavelengths (reds). Another example along these lines is fresh snow, which is white precisely because it is composed of flakes of sufficient size to reflect, without scattering or absorbing, the entire visible spectrum of the incident light from the Sun. As snow is compacted and transforms into slush, it becomes grayish because it absorbs a significant fraction of the incident light. The colors of the atmospheres of the various planets of the solar system, which are irradiated by the Sun just as is Earth, thus reflect the composition of each atmosphere. The terrestrial atmosphere is blue (and it has become common to refer to Earth as the blue planet), and the Jovian atmosphere is pink. However, the Martian atmosphere, which is composed essentially of nitrogen, is rather rust colored because it carries a large concentration of dust, and that of Uranus is blue green, no doubt because of the presence of methane, which absorbs in the red.

The variations in the colors of the sunset are due to the lower atmosphere (essentially the troposphere) serving as a prism through which passes the last rays of sunlight. These rays, which light up the sky close to the horizon, arrive at Earth almost tangent to the surface of Earth and are strongly refracted[15] when they enter the atmosphere, thereby bending their trajectories as a function of their wavelength. The index of refraction of the atmosphere, which determines the degree with which the rays are bent upon entering the atmosphere, varies with wavelength, being larger for blues than for reds. Thus, red light rays are least affected upon entering the atmosphere and are transmitted to form the sunset, whereas blue light rays are bent and sent out of the picture. Refraction thus explains why, in mountainous regions, north-west-facing slopes, which are seen when looking south-east, become red in the evening before going dark: they are illuminated by the predominantly red light that is not refracted away by the atmosphere. Looking north-west, however (fig. 1.5), results in light being scattered onto the retina by the north-western sky; this light is predominantly in the red and yellow part of the spectrum, making for the splendid sunsets that have been caught by countless cameras over the years.

---

15 Light is refracted everywhere where the index of refraction varies, be it in a continuous fashion such as when the temperature varies (for example, in the air above hot desert sand, which leads to mirages), or simply above a heater, or at the interface between two different media (e.g., air and water). The latter case explains why a stick that pierces the surface of water appears bent where it enters the water. In high-school physics we learn that, even in a homogeneous medium, the index of refraction varies as a function of wavelength. This is done by experimenting with a prism, which transforms a white beam of light into a fan of colors.

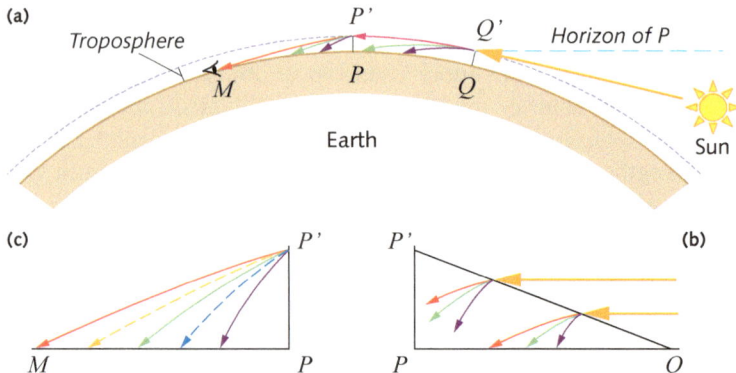

Figure 1.5 - (a) Schematic representation of global refraction of light in the troposphere at sunset. The portion $PP'QQ'$ of the troposphere is analogous to a prism, with the overall effect being to favor the transmission of red light into the slice PP'. **(b)** A zoom-in view of the prismatic region $PP'Q$, showing in detail two white beams penetrating the troposphere and being refracted in different directions depending on the color of the individual rays. **(c)** Observer M on the ground still sees the slice PP' of the troposphere, despite the Sun being below M's horizon, but the colors observed are mostly reds because the blues and violets no longer reach the plane PP'. Without this double refraction in the prisms $PP'Q$ and $PP'M$, no light from the Sun, which is under M's horizon, would reach this observer.

In the same way, the first dawn rays of the Sun that illuminate the eastern horizon are from the longer-wavelength part of the spectrum—the reds. This phenomenon is shown in figure 1.3, where we see the reddish troposphere in the morning with a blue stratosphere above sandwiching a blinding Sun in between. The reddish rays forming this image were bent upon entering the atmosphere, forming an image that would be impossible to form had the rays travelled in straight lines. A remarkable example of this effect is shown in figure 1.6, where the mountains of Corsica are visible over the pine trees along the Mediterranean coast at Nice, as if the peaks reach heights of over 4000 m. Rays travelling in a straight line from these mountains never reach Nice, being blocked by the curvature of the Earth. The red glow may be seen a good hour before the sky becomes blue, after having transitioned slowly through roses and pinks.

Figure 1.6 - The mountains of Corsica seen in the early morning from the Mediterranean coast at Nice. The mountains are visible because the red wavelengths are curved due to refraction in the atmosphere. Later in the day, when the sky has become blue, the mountains are no longer visible.

Sources of light other than the Sun are present in the low atmosphere, such as all the streetlights that illuminate cities at night. On a calm night, the light striking our retina from nearby streetlights (say, at less than 200 m) is very stable. However, light from more distant streetlights (say, over a kilometer away) seems to fluctuate. For example, from the coastal boardwalk at Nice (the *Promenade des Anglais*), once night has fallen, nearby streetlights can be seen as well as distant streetlights, far away around the contour of the Bay of Angels. Light from the former is free of fluctuations, whereas light from the latter, which are identical to the former, constantly twinkles, as if the distant streetlights were being constantly shaken. This twinkling is due to the random motion of molecules in the air between the observer and the distant streetlights. Given sufficient distance, scattering of light by these constantly moving molecules becomes sufficient to deviate the light in a fluctuating fashion. This effect is made possible only because the mean free path of molecules is of the order of the wavelength of the light; that is, from about 0.50 to 0.78 µm.

4. Heat exchange through the atmosphere

To complement the discussion above of the temperature distribution in the homosphere, we consider now the thermal exchanges within the atmosphere—a subject of great importance in the discussions of global warming. The flux of solar radiation toward the disk presented by Earth,[16] called the *solar irradiance*, is about 1361 W m^{-2} (watts per square meter). This magnitude varies little on the scale of decades or centuries, which explains why it is often called the *solar constant*. However, on the scale of the variations in Earth's orbit around the Sun[17] (of the order of 20 000 to 100 000 years), these variations become large enough to cause variations in temperature of the order of 20 °C, which explains the ice ages and the interglacial retreat of the glaciers.

To estimate the rate at which Earth is heated by solar radiation, two corrections must first be taken into account. First, the albedo[18] must be subtracted from the incident radiation. The albedo is the fraction of energy reflected (essentially by clouds, snow-covered surfaces, and the oceans) back toward space. This energy does not contribute to heating the planet. For Earth, the overall albedo is roughly 30%, which reduces the net irradiance received by Earth from the Sun to about

---

16 Seen from the Sun, or from infinitely far away, Earth, which has a radius R, appears to be a flat disc of area πR^2, whereas the real area of the terrestrial sphere is $4\,\pi R^2$. Each square meter of surface thus receives, on average, one quarter of the irradiance directed toward Earth. However, the various square meters do not receive their dose at the same time.

17 The periodic variations in the orbit of Earth about the Sun, which, when combined with the precession of the axis of Earth, are at the origin of the alternating ice ages and interglacial periods, are known as MILANKOVITCH cycles in honor of the Serbian geophysicists Milutin MILANKOVITCH (1879–1958).

18 The word albedo comes from the Latin *albus* (white), as do words such as alabaster or albino.

953 W m^{-2}. Second, because Earth is not a disc with area πR^2 but rather a sphere with area $4\pi R^2$, the net irradiance must still be divided by four to get a good estimate of the average power received on the surface of Earth per square meter. The result is about 240 W m^{-2}.

We shall now attempt to deduce a meaningful approximation of the average temperature at the surface [19] of Earth. Assume first that the thermal equilibrium of the planet implies that its own infrared radiation toward space exactly balances the incoming solar irradiance of about 240 W m^{-2}. To deduce the average temperature of Earth, we use the STEFAN-BOLTZMANN's law (see insert I1.4), which expresses the flux emitted as a function of the temperature of the emitting body. This leads to a temperature at the surface of 255 K, or about −18 °C. Clearly, this simple calculation is insufficient for understanding the average temperature of our planet, which hovers around 15 °C rather than −18 °C. Some mechanism must therefore exist that limits the cooling of our planet by reradiation into space. Neither conduction nor convection of heat within the atmosphere may be invoked because, although they transfer heat from one point to another, their overall effect on the temperature of Earth is zero. The mechanism responsible for the gap between −18 °C and 15 °C is the greenhouse effect, which is often invoked during discussions of global warming. This effect certainly justifies dedicating a paragraph and an insert. On the scale of the atmosphere, this effect is analogous to what happens in garden greenhouses or in glassed-in verandas. These structures reflect back toward their interiors a significant fraction of the infrared radiation emitted from the inside by the ground, persons, furniture, or plants, thereby reducing their cooling.

Just as for the blue sky, the origin of the greenhouse effect is found within the mechanisms of interaction between light and the molecules in the atmosphere. Its explication requires notions of quantum physics and is summarized for readers with more advanced knowledge in insert I1.5. In this paragraph, we limit ourselves to demonstrating its existence and quantifying it. Unlike the Sun, Earth is much too cold to emit visible radiation—it radiates heat toward space instead in the infrared and far-infrared spectral ranges, which is between 1 and 100 µm. Starting from a realistic estimate of the average temperature of Earth's surface of 288 K, or 15 °C, STEFAN's law tells us that the surface of Earth must emit 390 W m^{-2} instead of the 240 W m^{-2} received. As indicated in figure 1.7, the atmosphere must clearly intercept and return toward the surface the difference: 150 W m^{-2}. The greenhouse effect is undeniable; without it our planet would be very cold and quite inhospitable.

19　The very notion of temperature runs into several difficulties: because temperatures do not add as do intensities or flows, defining or calculating an average is not simple. Nevertheless, we accept that the average of temperatures measured at diverse points around the globe is sufficient to characterize a global temperature for the planet.

I1.4 - Thermal radiation, conduction, and convection

All surfaces emit electromagnetic waves, an example of which is visible light. The surface is said to radiate. The energy carried by these waves is intercepted by all media in the line of sight of the emitting surface. In turn, these media can reflect, absorb, or transmit this energy. So-called black surfaces are those that absorb all the radiation they receive. All surfaces emit radiation whose brilliance b varies as a function of the wavelength λ and of the temperature T according to PLANCK's law $b(\lambda, T) = A/\lambda^5(e^{c/\lambda T} - 1)$, where the constants $A = 1.1927 \times 10^{-16}$ W m^2 and $c = 1.44 \times 10^{-2}$ m K. The spectral distribution given by this law shows that warm bodies such as the sun, stars, or flames, can emit visible light, but that cold bodies emit in the infrared, which is outside the visible range. Overall, by summing up the contributions of each wavelength, the power emitted per unit surface area of a blackbody follows the STEFAN-BOLTZMANN's law $P_0 = \sigma T^4$, where $\sigma = 5.674 \times 10^{-8}$ W m^{-2} K^{-4} is the STEFAN-BOLTZMANN's constant.

Numerous bodies, such as Earth, act as black bodies, but only in the infrared region of the spectrum. The STEFAN-BOLTZMANN's law for such bodies is written as $P = \varepsilon \sigma T^4$, where ε is a coefficient less than or equal to unity called the *emissivity*. In the case of the terrestrial surface, which is at a temperature sufficiently low to emit only in the infrared, we can assume $\varepsilon = 1$. In general, to calculate thermal exchanges by radiation between surfaces, their relative orientations must be considered. However, for the exchange between Earth and space, which is perfectly visible, this optical effect provides no correction. The only correction to the STEFAN-BOLTZMANN's law comes from the fact that the medium traversed (i.e., the atmosphere), is not perfectly transparent but absorbs a part of the radiation, thus preventing it from reaching space.

Independent of radiation, every layer of matter with a temperature difference between its surfaces conducts a flux of heat between them that is proportional to the temperature difference between the surfaces and to the thermal conductivity of the material. The flux φ per unit area crossing a plane parallel to the surfaces, in units of W m^{-2}, is a function of the local temperature gradient T, as expressed by FOURIER's law $\varphi = -\kappa\, \mathbf{n} \cdot \nabla$, where κ is the thermal conductivity of the medium. For a dilute gas such as air, the conductivity is due to the motion of the molecules and can be deduced from the kinetic theory of gas. It is relatively small near the ground ($\kappa \approx 1$ W m^{-2} K^{-1}), but can attain magnitudes 10 times greater for water and 1000 times greater for metals.

Convection is a mechanism that amplifies the exchange of heat by conduction by constantly replacing warm fluid with cold fluid, which then needs to be warmed itself and replaced. Analyzing complex convection requires a knowledge of the flow, including fluctuations and turbulence. In practice, approximations based on experiment are often used. In general, the net result of convection is to multiply the flux due to conduction alone by a coefficient that can reach values of 10 to 1000, which represents a significant change in the order of magnitude.

I1.5 – On the greenhouse effect in the atmosphere

When a photon encounters a gas molecule, the latter can capture the energy quantum of the photon and translate it into a quantum of molecular vibration in such a way that energy is conserved. But this must be done without changing the magnitude of a quantity characteristic of each molecule: its electric dipole moment. What is the electric dipole moment? Every molecule contains a group of positive electric charges and a group of negative electric charges. Each group has a barycenter associated with it and, if the two barycenters are not coincident, the molecule resembles an electric dipole, which is characterized by its electric dipole moment.

When the barycenters are in the same place, such as in the symmetric diatomic molecules N_2 and O_2, the electric dipole moment is zero. The requirement of invariance prevents symmetric diatomic molecules from exchanging any energy with photons. When molecules such as these, which can oscillate only by stretching and compressing along their axis, are exposed to infrared radiation, their dipole moment, which is zero before the encounter, must remain zero after. Therefore, these molecules cannot capture photons; neither those emitted by the sun, nor those reflected from the surface of Earth. They are thus transparent to the corresponding electromagnetic radiation.

In contrast, more complex molecules such as H_2O, CO_2, or CH_4, offer other vibrational modes in which to store energy. For H_2O, the two hydrogen atoms can vibrate around an average position by beating like wings on either side of the oxygen atom. For CO_2, which is a linear molecule with double bonds between the carbon atom and each oxygen atom, the distribution of peripheral electrons and the minimization of their mutual repulsion is what stores the energy. These various modes allow these molecules to store energy from photons while still conserving their electric dipole moment. Other molecules, such as nitrous oxide N_2O and ozone O_3, also possess vibrational modes that allow them to absorb energy from photons carrying the requisite energy.

The fact that this transfer of energy to certain molecules can only occur for certain quanta of energy implies another important condition: the energy of the photon must be comparable to that of the vibrational mode of the molecule. This explains why these molecules are transparent to visible light from the Sun and why, within the infrared spectrum emitted by Earth, each species such as H_2O, CO_2, or O_3, exhibits absorption bands, as indicated in the figure below, with transparent bands in between. This results in the significant variations between their contributions to the greenhouse effect. Recall, for example, that the contribution of water vapor, per unit mass, is 6 times greater than that of carbon dioxide, and that of methane is 21 times greater.

In the homosphere, the molecules that capture part of the infrared radiation emitted by the ground and the oceans exchange it through collisions with all surrounding molecules, thereby redistributing the energy to the ensemble, as per the ideas presented in insert I1.3 on the kinetic theory of gas. The consequence is clear: an increase in the overall kinetic energy of the molecules in the gaseous mixture implies an increase in the temperature of the gas and, as per the STEFAN-BOLTZMANN's law, and an increase in the infrared radiation emitted by each fraction of the gas. The troposphere and the stratosphere thus receive some energy, which they store in the form of heat and reemit by radiation both toward space and toward the ground. The fraction that goes toward space is added to the direct radiation from the surface that the troposphere has not intercepted to give the overall outward flux of 240 W m^{-2} emitted by the planet toward space. The fraction that returns toward the surface constitutes the greenhouse effect, which has an average intensity of 150 W m^{-2}.

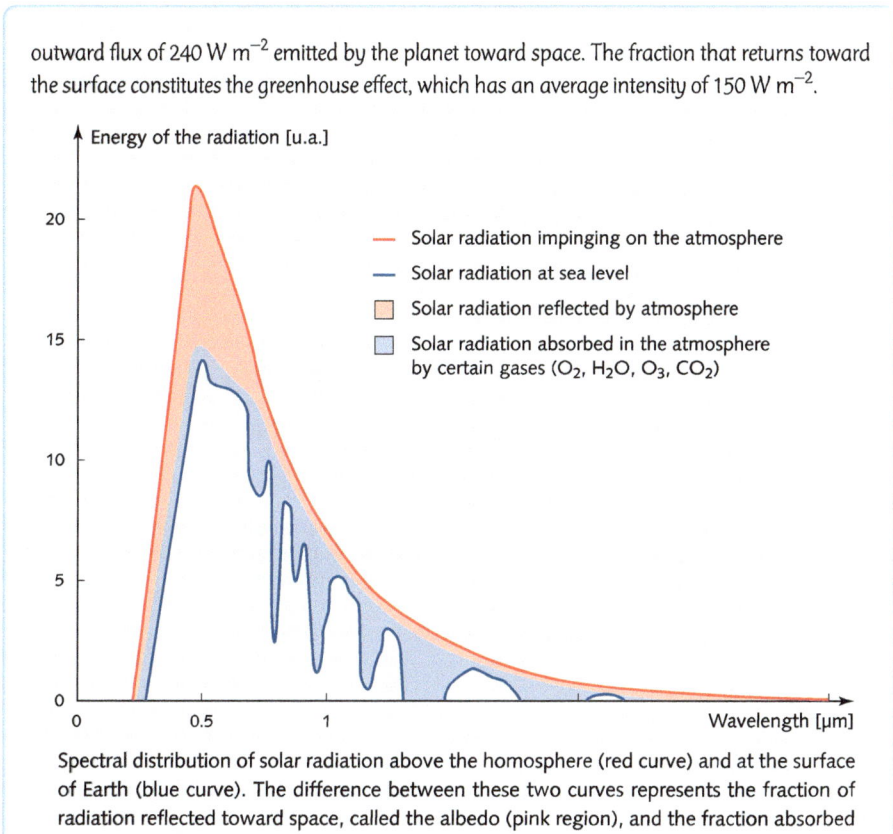

Spectral distribution of solar radiation above the homosphere (red curve) and at the surface of Earth (blue curve). The difference between these two curves represents the fraction of radiation reflected toward space, called the albedo (pink region), and the fraction absorbed by the main greenhouse gasses (blue region), for which the absorption bands are clearly seen.

The questions currently debated by scientists striving to understand the mechanisms that cause variations in the climate and to model them are not the existence of the greenhouse effect or the warming observed over the course of the 20th century. These questions fall well beyond the scope of this book but are treated in specialized texts.[20] What is the real extent of this warming? How can we evaluate, compare, and model the contributions to the greenhouse effect[21] of gasses such as water vapor, carbon dioxide, or methane? What is their relative contribution to the warming? What could be the direct contribution to this warming of the variations in solar activity? And what are the consequences of fluctuations in Earth's magnetic

---

20 Those interested by these questions are referred to the books of Jean JOUZEL, Claude LORIUS, and Dominique RAYNAUD, *Planète blanche, les glaces, le climat et l'environnement*, Odile Jacob, 2008; André LEGENDRE, *L'homme est-il responsable du réchauffement climatique?*, EDP Sciences, 2009; R. DAUTRAY and J. LESOURNE, *L'humanité face au changement climatique*, Odile Jacob, 2009.

21 Some elements can be found in insert I1.5.

field, which themselves lead to fluctuations in the flux of high-energy particles coming from the Sun (i.e., cosmic rays)? Above all, because human activity continues to transform fossil fuels containing carbonaceous materials (coal, petroleum, and gas) into carbon dioxide, by how much is this greenhouse effect going to strengthen and at what rate is Earth going to continue to warm? The history of the climate, marked by long ice ages [22] interrupted by remissions such as the Holocene in which we find ourselves now, is fairly well known and indicates that the two temperatures of −18 °C and 15 °C that come from the simple calculations outlined above define rather well the range of possible temperature variations.

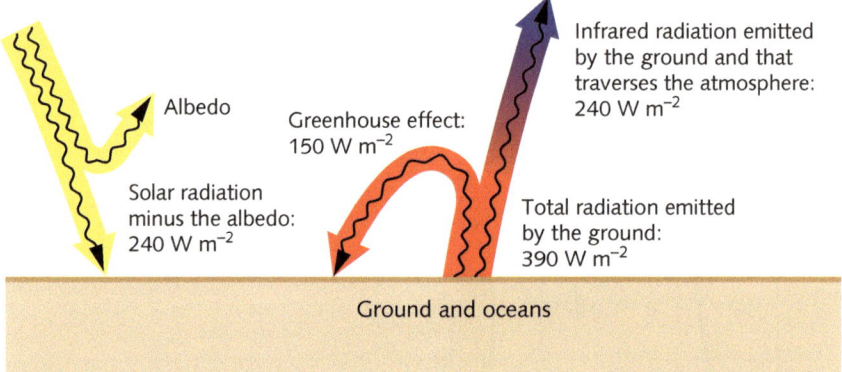

Figure 1.7 - Main contributions to the thermal equilibrium of Earth. The greenhouse effect accounts for the flux coming from the surface of Earth (150 W m^{-2}), which is intercepted by the atmosphere and sent back toward the ground.

Conclusion

Since its formation about 4.6 billion years ago, our planet has been marked by impressive evolutionary developments that have allowed life to develop. These changes progressively decreased the fraction of carbon dioxide in the atmosphere, initially near 98%, to its actual value close to 0.034%, and increased the fraction of oxygen, initially at zero, to its actual level close to 21%. This evolution allowed life to develop, led to the advent of *homo sapiens*, and produced all the forms of life that we know. The same evolution continues today, at its own pace; its main stages occurred long before modern man developed his industry. If worries exist about the future of the atmosphere and of life on Earth, which we will discuss more

22 Four glacial periods have occurred since the quaternary era. These are the Günz period, which ended 540000 years ago, the Mindel period, which ended 450000 years ago, the Riss period, which ended 180000 years ago, and the Würm period, which ended only 16000 years ago. Each period lasted about 60000 years, except for the Würm period, which seems to have lasted some 100000 years. These changes in the climate, called *MILANKOVITCH cycles*, predate human activity and are explained by variations in Earth's orbit and by its precession on its axis of rotation.

in the epilog, we can only say that it is not the planet that is threatened, but rather all the current forms of life that inhabit it.

The atmosphere is an open system between the ground and space and is subject to extremely powerful phenomena. All that happens at the surface of the Sun, be it variations in the number of sunspots or ejections of matter, has repercussions on our atmosphere. Although these are hard to model with precision, we know that the magnetosphere and the atmosphere attenuate them to the point that life on Earth is relatively well protected. In addition, all major incidents on the planet itself, such as industrial catastrophes, wars, volcanic eruptions, or the massive injection of carbon dioxide during the industrial era, significantly disturb the atmosphere, which spreads the effects over all the continents and over long periods of time.

Chapter 2

The atmosphere in movement

You think, then, that aerostatic science has said its last word?

Not at all! Not at all! But we must look for another point in the case, and if we cannot manage to guide our balloon, we must at least try to keep it in favorable aerial currents.

(Jules VERNE, *Five Weeks in a Balloon*)

© Springer International Publishing AG 2017
R. Moreau, *Air and Water*,
DOI 10.1007/978-3-319-65215-3_2

L et us reexamine this atmosphere, presented as calm and immobile in the first
chapter but in reality in constant movement over a very large range of scales.
Some winds are relatively stable and well established; others appear quite random
and are the source of serious conundrums for meteorologists. A relatively methodical observation thus appears necessary to extract the main properties: we shall
begin with the largest scales; namely, those with horizontal dimensions of the same
order of magnitude as the radius of Earth (6730 km) and that encompass the entire
thickness of the troposphere (8 to 15 km). The circulation of air on these scales in
fact structures movement within the atmosphere on the whole. We shall see later
that oscillations on more modest scales can explain seasonal phenomena, such as
monsoons, or random phenomena, such as the back and forth between high- and
low-pressure zones. On scales approaching 500 to 1000 km, we shall focus on the
mechanisms that drive significant meteorological phenomena, such as cyclones
and low-pressure zones, which are remarkable in their rather precise organization. On smaller scales, we shall examine periodic spatial fluctuations, such as lee
waves, or periodic as in the diurnal thermal winds. However, the violent intermittent events that often occur in the atmosphere, such as thunderstorms, tornadoes,
or the whims of precipitation, are reserved for the next chapter.

1. Large-scale circulation within the atmosphere

1.1. The trade winds, the HADLEY cell, and the subtropical jet stream

The driving force behind the circulation within the atmosphere is sunshine, which
reaches its maximum where the Sun is at its zenith. This zone is centered over the
equator only during the equinoxes (March and September), whereas it reaches the
tropics at latitudes of ±23° 26′ during the summer solstices (June in the Northern
Hemisphere and December in the Southern Hemisphere). Note that, because Earth
rotates once about its axis every 24 hours, this zone of maximum sunshine moves
rapidly toward the west: its speed with respect to an observer on the ground is
approximately 1600 km h^{-1}. It is also fairly vast: the diameter of the circle where
solar radiation is 90% of its maximum is about 5000 km. Within this zone, the
ground is heated intensely by the radiation and, in turn, it heats the atmosphere by
conduction and convection.

The air above this enormous fast-moving circle is thus lighter than the surrounding
air, so that the net result of weight and ARCHIMEDES' buoyancy lifts it, systematically
creating a local low-pressure zone that draws a horizontal flow of air from the

surrounding atmosphere. Air flowing in from the north moves southward, and the CORIOLIS force (see insert I2.1) deviates it toward to the right, which is west. The air flowing in from the south moves northward and the CORIOLIS force deviates it to the left, which is also west. The air to the west of this circle is immediately absorbed by the arrival of the warmed air. The air to the east, which has already been heated and carried westward over the preceding hours, also participates, although rather feebly, to this westward wind.

Figure 2.1 schematically represents these air currents, which constitute the trade winds. These are the winds near the equator that blow west and that are well known to the sailors that they push westward. These reliable winds, although of modest speed of the order of 20 km h^{-1}, pushed the schooners of Christopher COLUMBUS towards the Caribbean and South America and those of Hernán CORTÉS towards Mexico. At sea level, an extremely tranquil zone develops where the trade winds coming from the north meet those from the south. This area, feared by sailors partaking in the international around-the-world regattas, has earned the name the *doldrums*.

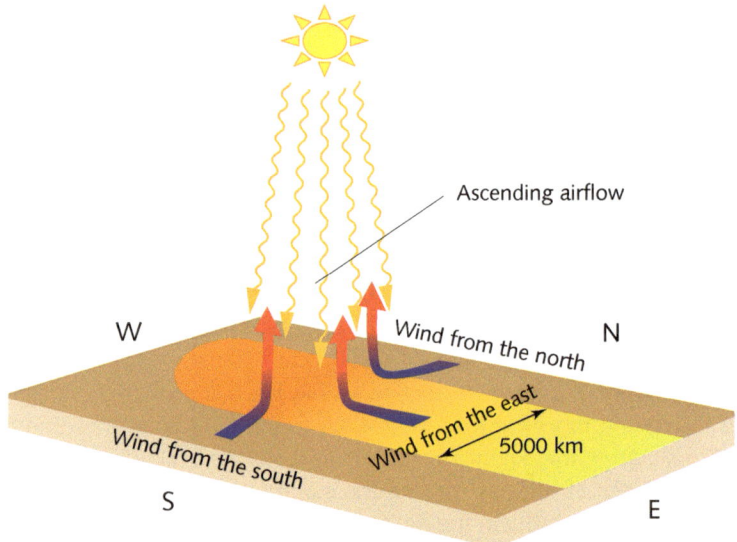

Figure 2.1- Schematic illustration of the mechanism driving the trade winds, of the ascending movement resulting from the reduced density of the overheated air, and of the westward airflow at the equator. The diameter of the overheated zone is of the order of 5000 km and moves westward at about 1600 km h^{-1} at sea level. The speed of the resulting wind is much less; of the order of 20 km h^{-1}.

Overheating of the zone over which the Sun is at the zenith also results in an ascending air current, which transports the thus-lightened air to high altitudes. This air becomes laden with moisture when it passes over the oceans and the deep equatorial vegetation. It cools in relation to its altitude and progressively decompresses, leading to condensation of its water vapor and the formation of large clouds called *cumulonimbus clouds*. These clouds are typical of the tropical regions and, near the equator, their summits can reach as high as the tropopause, about 15 km above sea level. The organization of these cloud systems varies with longitude because the continents, which are relatively dry and cold at night, release less moisture than the oceans. These differences lead to an intense mixing of these masses of air with differing temperature and water content. When the warmer and lighter masses of air pass above the heavier masses, the sheared regions between them form what are called *warm fronts*. *Cold fronts*, on the contrary, occur when a cold, heavy mass of air passes under a warm mass of air. In the tropical regions, the end of the day brings warm fronts fairly systematically onto the western coasts of the continents, which find their way by daybreak to the eastern coasts. Cold fronts are found on the opposite coasts: east coast at the end of the day, and west coast at daybreak. Both fronts can bring precipitation and thunderstorms, which are quite typical of the tropical regions.

Once at the tropopause, the ascending air current, dried and cooled, has lost is upward momentum due to its struggle against its own weight. It can progress no further and cannot penetrate the stable layer that covers the troposphere like a lid. However, its mass flow is conserved, which forces it to move toward the north in the Northern Hemisphere or toward the south in the Southern Hemisphere. At this point, the CORIOLIS force comes into play, orienting these movements toward the west in both hemispheres. From Earth, we thus see in each hemisphere the formation of a westward current, which is called the *subtropical jet stream*. The angular momentum[1] of this current is conserved (see insert I2.2) because no significant force counters it at these altitudes, where viscous friction is negligible. The altitude of the jet stream is much greater than the thickness of the atmospheric boundary layer, where friction is localized (see inserts I2.3 and I2.4). Conservation of angular momentum is enforced on all mechanical systems that rotate and are free of friction, which endows this westward air current with two important properties.

On the one hand, this explains the high speed, of the order of 80 to 100 km h^{-1}, of this high-altitude wind. At the beginning of its ascension, the angular speed ω of the air from the equatorial regions is roughly that of the neighboring air at these low altitudes. To maintain its angular momentum ωr^2, its angular speed ω must increase as it gains altitude and as its distance r from the axis of Earth diminishes, forcing this air to turn ever faster than the planet. In fact, as latitude increases, r^2 decreases and ω increases to compensate because the product ωr^2 must remain constant.

---

[1] The definition of angular momentum and the laws of mechanics that govern rotating bodies appear in insert I2.2, which requires a physics knowledge at the level of introductory undergraduate physics.

I2.1 - Force, acceleration, and NEWTON's second law of motion

The equilibrium and motion of bodies obey well-established laws that themselves are founded on precise concepts. The first concept to introduce is mass, which represents the quantity of matter in an object. That our planet cause an acceleration g due to gravity is a reflection of the fact that each object of mass m is attracted toward the center of Earth by a force mg, which we call its *weight*. Although mass is an intrinsic property of each object, the same is not true for gravity, which varies slightly from point to point on Earth and quite strongly from one celestial object to another. Thus, an object's weight is not invariant, as was verified when Niels ARMSTRONG et Buzz ALDRIN set foot on the moon on July 20, 1969.

Objects are also subjected to other forces, such as the push exerted on their surface by a surrounding fluid, which expresses itself locally as the product of pressure p and the surface area over which the pressure is applied. For airplanes or birds that can fly, we have to separate this force into two components: the lift and the drag. For automobiles, the reaction to the ground is combined with the lift, and air drag leads to the familiar coefficient C_x. For all objects at rest and immersed in a fluid, the lift is given by the buoyancy of ARCHIMEDES, which equals the weight of the fluid displaced by the object.

NEWTON's second law of motion demands that, for each part of a body of mass m moving at velocity $\mathbf{V}$, the sum $\mathbf{F}$ of applied forces equals the product $m\gamma$, where $\gamma = d\mathbf{V}/dt$ is the acceleration of this part of the body and t is time. Note that this law holds only in an inertial reference frame.

When a body moves with an angular velocity ω along a trajectory whose local radius of curvature is r, it undergoes a centrifugal acceleration of magnitude $\omega^2 r$ directed radially away from the center of the curve. The common practice is to associate with this acceleration a fictitious force $m\omega^2 r$, which is called the *centrifugal force*. Cyclists and motorcycle riders compensate for the centrifugal force by leaning toward the inside of the curve, thus using their weight to balance the torques on their body and machine. On Earth, and especially in the atmosphere and in the oceans, all trajectories are curved so the centrifugal force is inevitable.

Moreover, the moment we consider describing the movement of air or water as can be observed on Earth, it is convenient to write the fundamental law of dynamics for the frame of reference of Earth, which is not an inertial frame because it is accelerating. Thus, another very important fictitious force enters the equation called the CORIOLIS force[1]. With the help of the figure below, which represents a disc rotating with angular velocity Ω (the direction in which Earth rotates when viewed from above the North Pole), we shall satisfy ourselves here to giving a qualitative interpretation of this force. An object is launched from the center O toward point M, which is on the periphery of the disc. An external fixed (or inertial) observer, wherever she might be in the space around the disc, will see the object travel in a straight line from O to M. However, the disc turns constantly to the left of the trajectory, so an observer fixed to the disc sees the object's trajectory curve to the right of the line OM and arrive at the point M'. At the same time, for the inertial observer, the point M' has arrived at original position of point M after travelling $MM' = \Omega t$.

---

1   This fictitious force was discovered by Gaspard-Gustave CORIOLIS (1792–1843), a French mathematician and engineer and a professor at the *Ecole Polytechnique* and the *Ecole Centrale des Arts et Manufactures* in Paris.

Illustration of the CORIOLIS force in the simple case of a disc rotating with angular velocity Ω. The blue curve shows the rightward-curving trajectory seen by an observer fixed to the disc. The line OM is the trajectory seen by an observer in an inertial frame of reference.

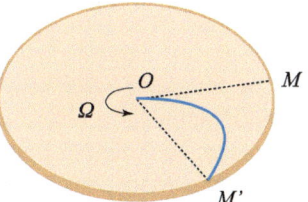

What we shall retain from this discussion is that, as seen from Earth, the CORIOLIS force bends trajectories to the right in the Northern Hemisphere and to the left in the Southern Hemisphere. Note also the order of magnitude of the deviation $MM' = \Omega rt$. The rotational speed of Earth is about 10^{-4} rad s^{-1} (one revolution in 24 h) so, if the trajectory takes approximately one minute to complete, such as for emptying a sink, the deviation is about less than 10^{-3} revolutions. This is negligible for emptying a sink and cannot explain the characteristic vortex that makes several revolutions while the sink empties. However, if the duration of the trajectory (at least 5 h in the case of atmospheric low-pressure zones) is comparable to the time for one revolution of Earth (24 h), then the CORIOLIS force cannot be neglected.

So what then is the origin of the rotational flow upon emptying a sink? This is the net result of shear forces that act at the interfacial layer of the fluid in contact with the surface of the sink. The properties of this interfacial layer can be better appreciated with the help of insert I2.3. There we shall find that the fluid particles close to the sink surface rotate while flowing and we can understand that, overall, the rotation gathered in the drain can lead to an overall rotation that gives the well-known vortex.

On the other hand, the component of the velocity oriented toward the poles decays rather rapidly, this time due to conservation of kinetic energy, in such a way as to compensate for the increase of the component of ωr directed along the longitudinal lines. In fact, because ωr^2 remains constant, when a mass of air moves away from the equator, the distance r to the axis diminishes and, as a result, ωr increases. Thus, this dry air has no choice but to dive back toward the ground. In so doing, it becomes more and more compressed, so that, in keeping with the equation of state for an ideal gas, its temperature increases. Arriving at sea level at latitudes of ±30° to 35°, this hot dry air creates the well-known belt of deserts in these regions, such as the Sahara and the Gobi desert in the Northern Hemisphere, or the Australian desert and the high Andean plateaus in the Southern Hemisphere.

We can thus understand why the two helical coils in figure 2.2, which are characteristic of the tropical regions, loop back upon themselves. The projection of this circulation onto a meridional plane is known as the HADLEY cell in honor of the English lawyer and amateur meteorologist George HADLEY, who was the first to suggest their presence and structure in 1735. The double-helical coil represents the combination of the cell with the westward equatorial airflow at the ground and the high-altitude subtropical jet stream. It would be wrong to think of this circulation as stable. While the westward equatorial airflow near the ground is slow and relatively stable, the jet stream, on the contrary, is rapid and agitated. It is characterized both by long-wavelength oscillations, visible in figure 2.3, and by a high level of turbulence. In addition, because the HADLEY cell is centered on the zenith and not on the equator, the large helical currents shown in figure 2.2 are subject to seasonal displacements, as shown in figure 2.4. These are important and help explain phenomena such as monsoons, which we shall revisit later.

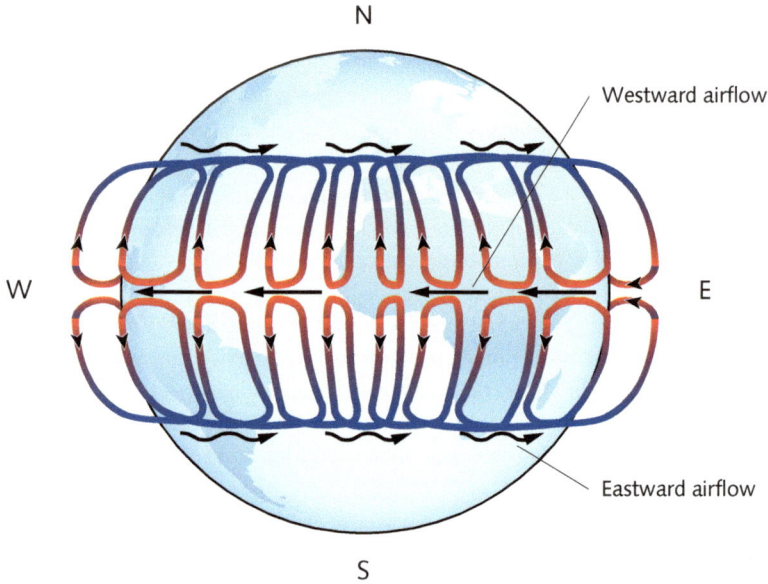

Figure 2.2 - Schematic view of the winds in the HADLEY cell during the equinoxes, when the cell is centered on the equator. The slow and stable westward airflow is represented by left-pointing arrows. The eastward airflow, or jet stream, is represented by wavy arrows and appears at higher altitudes in the tropical latitudes and is much faster and more unstable. These winds combine to form a double-helical circulation; strictly speaking, the HADLEY cell is the projection of this pattern onto the meridional plane.

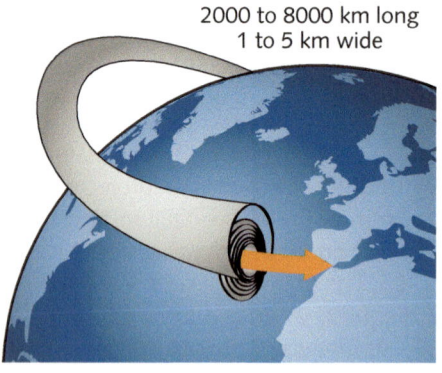

Figure 2.3 - Schematic representation of the position and dimensions of typical oscillations in the polar jet stream in the Northern Hemisphere. [© US Centennial of Flight Commission]

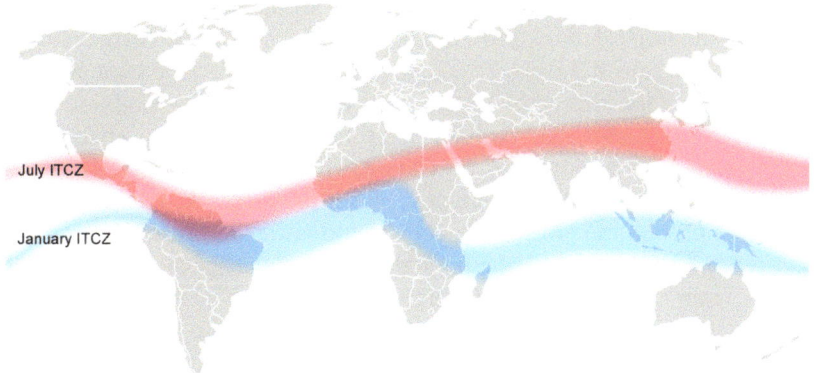

Figure 2.4 - Maximum seasonal variations in the westward equatorial air currents (ITCZ: *Inter Tropical Convergence Zone*; red indicates summer, blue indicates winter). [© Mats HALLDIN]

1.2. Polar cells, FERREL cell, and the polar jet stream

A convective airflow analogous to that of a HADLEY cell exists above each polar ice cap, or from 60° to 90° of latitude. However, it is much more strongly influenced by the CORIOLIS force than is the HADLEY cell. In fact, the angular velocity that enters the expression for the horizontal component of this force is the projection onto the local vertical direction of Earth's angular velocity with which Earth rotates (fig. 2.5).

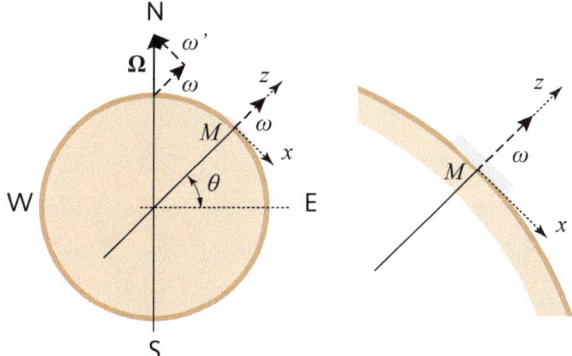

Figure 2.5 - Near point M at the latitude θ, in a local coordinate system (Mx, My, Mz), the components of the angular velocity of Earth are $[\omega' = -\Omega \cos\theta, 0, \omega = \Omega \sin\theta]$. Using $[u, v, 0]$ for the components of the horizontal velocity $\mathbf{V}$ in the local troposphere (represented in gray near point M), the components of the CORIOLIS acceleration $2\Omega \times \mathbf{V}$ are $[-2\omega v, 2\omega u, 2\omega' v]$. This shows that only ω, the component of Ω in the local vertical direction, contributes to the horizontal motion.

On the surface of our planet, the CORIOLIS force is thus strongest near the poles and weakest in the tropical regions (on the contrary, the centrifugal force is strongest at the equator and zero at the poles). Moreover, near the poles, the thinness of the troposphere (7 to 8 km instead of 15 km at the equator) and the more rapid variation in the distance to Earth's axis modifies significantly the quantitative parameters that determine the local atmospheric circulation. The air descending toward the poles like the air that, in the HADLEY cell, descends in the tropical regions, is thoroughly dried out at altitude and promotes the Arctic and Antarctic deserts. Contrary to what is observed in the HADLEY cell, the high-altitude airflow near the poles is particularly stable. It leads to near-permanent high-pressure zones in these regions, accompanied by very low temperatures.[2] Near the ground, it also leads to a fairly stable regime of south-west winds in the Arctic and north-west winds in the Antarctic.

In each hemisphere, between the HADLEY cell whose maximum latitude is roughly 35° and the polar cell that is situated above 60°, sits an intermediate region call the *FERREL cell*, named in honor of William FERREL, the American meteorologist of the late 20th century. This cell is distinguished essentially by the continuity required of the overall atmospheric circulation, which explains the direction of the currents in the meridional planes that are the inverse of those in the HADLEY cells and the polar cells. The overall global atmospheric circulation is composed of these three cells, as shown schematically in figure 2.6.

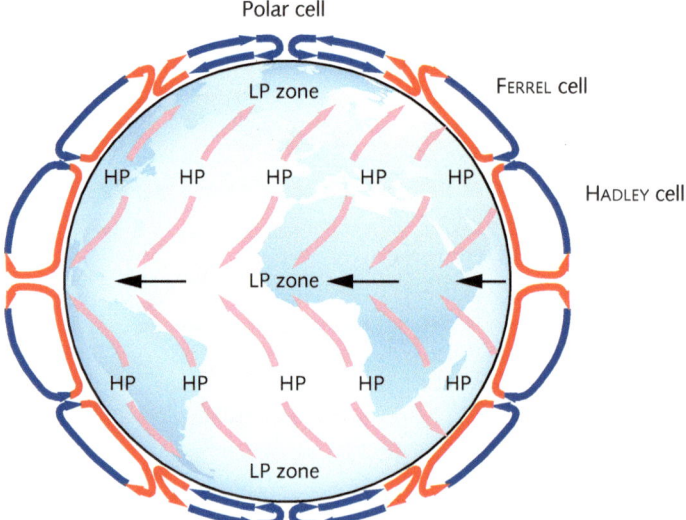

Figure 2.6 - Schematic representation of the global atmospheric circulation. LP marks relative low-pressure zones that result from the ascending airflow, HP marks two belts of high-pressure zones separated by intermittent low-pressure zones.

---

2   The record, recorded in Antarctica at the Russian station Vostok on July 21, 1983, is −89.2 °C.

Remarkably, at the top of the ascending zone located between the FERREL cell and the polar cell, another eastward airflow occurs, which is called the *polar jet stream*. This polar stream, shown in figure 2.3, flows even more rapidly than the subtropical jet stream, essentially because, at latitudes close to 60°, the CORIOLIS force exerts a much greater influence than in the tropical regions. In fact, its velocity can reach up to 300 km h^{-1}.

2. Low-pressure zones and cyclones

2.1. Formation and direction of rotation of atmospheric low-pressure zones

We have just seen that, on scales smaller than the radius of Earth, and particularly at the middle latitudes, fronts interspersed in time and space separate low-pressure zones from high-pressure zones. Wherever they may be and whatever their intensity, low-pressure zones are accompanied by curving winds; in other words, of movement relative to that of Earth. These winds are called *cyclonic*[3] because they turn in the same direction as Earth (counterclockwise as seen from above in the Northern Hemisphere and clockwise in the Southern Hemisphere). They are the result of a subtle and relatively unintuitive game between the forces at play, which we shall examine in several stages.

To start with, as indicated in insert I2.1, recall that all bodies traveling along a curved trajectory experience a centrifugal force. To balance it, a counteracting force (i.e., centripetal) must appear. This can only come from differences in pressure: the turning fluid medium inside a low-pressure zone pushes against the surrounding medium, creating at their interface a pressure that exceeds that within the rotating system. This is well known: the rotation induced in a cup of tea by a spoon deforms the free surface that is initially horizontal into a parabolic shape with a dip at the center of the cup. In reality, because the air on the outside of this depression is essentially at rest, the pressure external to the rotating system is governed by hydrostatics, discussed in chapter 1. Thus, it is the pressure internal to the system that is reduced. We can estimate the decrease in pressure by using BERNOULLI's law, also introduced in chapter 1, and which gives a reduction in pressure of the order of $\rho V^2/2$, where ρ denotes the air density and V is the typical wind speed within a low-pressure zone.

When the intensity of the low-pressure zone becomes sufficiently strong to cause serious damage, such as breaking tree branches or damaging roofs, it is called a *hurricane* (in the Atlantic or northeastern Pacific oceans), a *cyclone* (in the Indian

---

3   The word *cyclone* comes from the Greek word *kyklos* (κύκλος), which means circle.

and south Pacific oceans), or a *typhoon* (in the northwestern Pacific ocean) and is given a name and categorized[4] in terms of severity. In this text, we use the generic technical term *cyclone* for all such phenomena. Category 5 cyclones are the most devastating: trees are uprooted, buildings are severely damaged, and flooding is catastrophic. An example from this category is the cyclone KATRINA that destroyed a large part of the city of New Orleans in the United States in 2005. In contrast, regions external to the low-pressure zones or cyclones, where the air remains calm, are high-pressure zones (or anticyclones). The appellation *anticyclone* does not indicate that these systems rotate in the direction opposite that of cyclones. Rather, high-pressure systems do not rotate at all because the mechanism that drives the rotation of low-pressure zones, which is analyzed in the following pages, is not present. For now, the important point to remember is that the air in high-pressure zones remains essentially at rest with respect to Earth.

We focus first on the higher altitudes, above the atmospheric boundary layer where the friction with the ground or with the oceans may be neglected. In a low-pressure zone with horizontal dimensions of the order of thousands of kilometers, the CORIOLIS force must be considered (see insert I2.1). To be simultaneously perpendicular to the direction of the wind (which is horizontal) and perpendicular to the local vertical direction, the CORIOLIS force must be radial and must intervene to balance the difference between the centrifugal force and the net pressure. This combination of forces, illustrated in figure 2.7, constitutes the geostrophic equilibrium, which is an important ingredient for understanding atmospheric movements and therefore merits a more detailed discussion.

In the low-pressure zone shown schematically in figure 2.7b, the wind speed remains perpendicular to both the local vertical direction and to the radial force due to pressure (we have seen that this force comes from the difference in pressure between the external high-pressure zone and the low-pressure zone and balances the centrifugal force). The wind direction is essentially horizontal, but at 90° to the radius of the low-pressure zone. The wind is thus said to be *azimuthal*. This rotating flow thus forms a large, more-or-less circular vortex whose velocity is perpendicular to the driving forces created by pressure differences.

---

4   To allow the authorities to predict possible damages, low-pressure zones and cyclones are named and classified by their intensity. This classification system is called the SAFFIR-SIMPSON *hurricane scale* and is summarized below:
   - tropical depression: wind speed less than 62 km h^{-1},
   - tropical storm: wind speeds between 63 and 118 km h^{-1},
   - category 1 cyclone: wind speeds between 119 and 153 km h^{-1}, 1.5 m waves,
   - category 2 cyclone: wind speeds between 154 and 177 km h^{-1}, 2 m waves,
   - category 3 cyclone: wind speeds between 178 and 209 km h^{-1}, 3 m waves,
   - category 4 cyclone: wind speeds between 210 and 249 km h^{-1}, 4 to 5 m waves,
   - category 5 cyclone, catastrophic: wind speeds greater than 249 km h^{-1}.

(a)   (b)

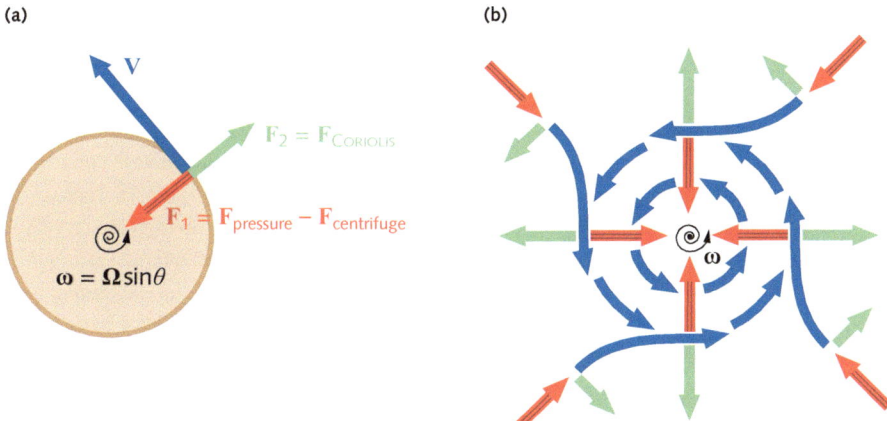

Figure 2.7 - Geostrophic equilibrium of an eddy in the horizontal plane in the Northern Hemisphere **(a)** Equilibrium between F_1 (the radial force resulting from the competition between the force due to centripetal pressure and the centrifugal force) and the Coriolis $F_2 = 2\omega \times V$, which is also radial but opposes F_1. Here, ω denotes the angular velocity of Earth (see fig. 2.5) and V is the local horizontal velocity. Equilibrium requires that V be oriented azimuthally, which causes the cyclonic rotation. **(b)** Overview of the streamlines of airflow (everywhere tangent to the blue velocity vectors) within a low-pressure zone, with the forces balanced.

This may seem paradoxical or even contrary to our intuition formed by observing flows on human scales, where fluid always flows from high pressure to low pressure or from high altitudes toward lower altitudes. But it is precisely this change in orientation of flows to 90° from the driving forces that distinguishes large-scale geophysical flows, which are dominated by the Coriolis acceleration, from small-scale flows (less than 100 km) on which it is not significant.

The large eddy is systematically cyclonic, which means that, in both hemispheres, it turns in the same direction as Earth about its axis. In other words, the moving air in such a cyclone turns faster than Earth does itself about the local vertical direction, which is inclined with respect to the axis of the planet. In effect, when observed from space instead of from the ground, these large atmospheric vortices all rotate in the same direction as Earth and represent a sort of over-rotation. In contrast, seen from the ground, the low-pressure zones in the Southern Hemisphere, where the latitude and the vertical component of the angular velocity of Earth have both changed sign, rotate clockwise. The observer is the one who is turned around; the direction of rotation in the inertial frame has not changed!

12.2 - Summary of mechanics of rotating bodies

Consider a body rotating about an axis $\varDelta$. To describe its motion, we use a cylindrical coordinate system (r, θ, z) centered on this axis and the unit vectors $(\mathbf{e}_r, \mathbf{e}_\theta, \mathbf{e}_z)$. Denote the angular velocity of the body by ω and the local velocity in the azimuthal direction θ by ωr. The momentum of a mass m moving with angular velocity ω is $m\omega r\,\mathbf{e}_\theta$ and its angular momentum with respect to the axis $\varDelta$ is $\mathbf{L}_\varDelta = m\omega r\,\mathbf{r} \times \mathbf{e}_\theta$. With respect to the axis $\varDelta$, the moment of inertia of this mass m is $I_\varDelta = mr^2$. If the body is subject to a force $\mathbf{F}$ creating a torque with respect to the axis is $\mathbf{M}_\varDelta = \mathbf{r} \times \mathbf{F}$, Newton's second law of mechanics implies that $d\mathbf{L}_\varDelta/dt = \mathbf{M}_\varDelta$. Note the similarities between the these quantities and the expressions describing rotating systems and systems undergoing arbitrary motion: the torque $\mathbf{M}_\varDelta$ replaces the force $\mathbf{F}$, the angular momentum $\mathbf{L}_\varDelta$ replaces the linear momentum $m\mathbf{V}$, and the moment of inertia $I_\varDelta$ replaces the mass m. Like mass, the moment of inertia represents the resistance of bodies to variation in their angular velocity. When no forces act, the motion of a rotating body is controlled by the invariance of angular momentum.

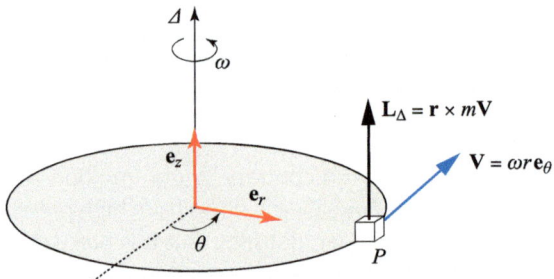

Velocity $\mathbf{V} = \omega r\,\mathbf{e}_\theta$ and angular momentum $\mathbf{L}_\varDelta = \mathbf{r} \times m\mathbf{V}$ for a point mass m rotating with angular speed ω about an axis $\varDelta$

One well-known illustration of the equality between the derivative of angular momentum and torque is the technique of figure skaters, who spread their arms and one leg to acquire a maximum of torque $\mathbf{M}_\varDelta$. Next, they retract their arms and leg to increase their angular speed by minimizing their moment of inertia. But this increase is only temporary because friction with the ice constantly eats away at their angular momentum.

Taking the dot product with velocity $\mathbf{V}$ of each side of the equation $d(m\mathbf{V})/dt = \mathbf{F}$, we obtain the general kinetic energy theorem with the classic result that the derivative with respect to time of the kinetic energy equals the power provided by external forces: $d(mV^2/2)/dt = \mathbf{F} \cdot \mathbf{V}$. For systems rotating about an axis, we obtain another form of the same result: $d(I_\varDelta\omega^2/2)/dt = \mathbf{M}_\varDelta \cdot (\omega r\,\mathbf{e}_\theta)$, and the kinetic energy of the motion is written $I_\varDelta\omega^2/2$.

The sign of the rotation of Earth is thus critical, and the moment has come to explain this dissymmetry between the two possible signs, already mentioned several times in the preceding pages. One may reasonably think that, at the moment at which the low-pressure zone forms, it contains an equal number of vortices of each sign and whose energy is of the same order of magnitude. Two mechanisms now come into play. First, within a cyclonic vortex, the effective angular velocity in an inertial reference frame is the sum of the angular velocity of Earth and the angular velocity of the low-pressure zone. This absolute velocity[5] is thus large and this vortex structure thus carries a large angular momentum, which guarantees it a long lifetime. In contrast, within an anticyclonic vortex, the absolute angular velocity is the difference between the angular velocity of Earth and that of the vortex, which means that this structure has small angular momentum and a correspondingly short lifetime. This is the first point that explains why, the moment they form, anticyclones become weaker than cyclones.

Next, anticyclonic vortices, which are still present although weakened, are rapidly stretched by the more powerful cyclonic vortices and progressively wrap around them (see appendix). In this way, they transforms into sheets or filamentous structures that rapidly lengthen. To conserve mass, this lengthening of the anticyclonic vortices must be accompanied by a reduction in their orthogonal dimension (see fig. A.15 in the appendix). Their cross section in the horizontal plane thus consists of ever longer filaments that wrap around the cyclonic vortices and become thinner and thinner. Viscous friction, which is proportional to the local shear forces or inversely proportional to the thickness of the sheared layer, is highly amplified by small-scale turbulence and thus manages to dissipate their energy into heat and thereby make them disappear.

We now turn to the lower altitudes, where friction becomes significant, even for large-scale flows. This lower portion of the troposphere, where friction is capable of perturbing the geostrophic equilibrium, is quite a special boundary layer because it is dominated by the CORIOLIS force; it is called the *EKMAN layer*.[6] Insert I2.3 introduces the concept of a boundary layer in a general context, and certain aspects of the atmospheric boundary layer are refined in insert I2.4. The EKMAN layer has two effects on the internal organization of low-pressure zones. As in all boundary layers, friction reduces the air velocity to zero at ground level and the velocity increases progressively with altitude until it reaches its value outside of the boundary layer, which satisfies the properties of geostrophic equilibrium. In addition, and this is quite specific to rotating fluids subjected to a significant CORIOLIS force, friction imposes a variation in the wind direction as a function of altitude. Here again is a curious property of flows affected by the CORIOLIS force that requires an explanation.

---

5   *Relative movements* means those movements that are relative to a reference frame that itself moves with respect to an inertial reference frame. The absolute velocity of a particle in an inertial reference frame is the vector sum of the velocity of the frame of reference and of the relative velocity of the particle. The same is not true for accelerations, whose combination gives rise to an extra term; namely, the *CORIOLIS acceleration*.

6   The properties of this boundary layer were discovered by the Swedish oceanographer Vagn Walfrid EKMAN (1874–1954) during his thesis studies in 1902.

To thoroughly understand this continuous variation in wind direction as a function of altitude, imagine a thin horizontal slice somewhere within the EKMAN layer, such as the block $ABCD$ shown in figure 2.8. The net friction on the upper and lower surfaces must balance the pressure forces and the CORIOLIS force discussed above. This would not be possible were the velocities V_+ and V_- oriented in the same direction. In fact, for the net friction from F_+ and F_- at all altitudes be directed radially, as for the CORIOLIS force, the line of action of these forces must not coincide, so that the sum of F_+ and F_- is directed orthogonal to their average direction. This imposes a continuous variation on the wind direction as a function of altitude. The result, illustrated in figure 2.9, is often represented by the projection onto the ground of the velocity profile. It is known as the EKMAN spiral.

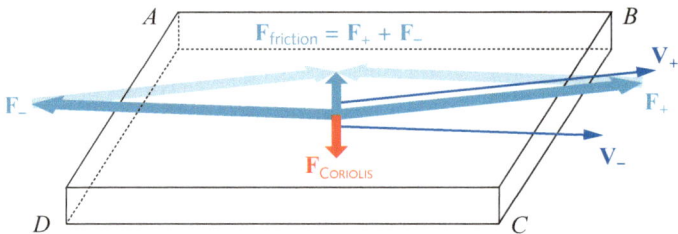

Figure 2.8 - The net force due to friction (blue arrows) that acts on the lower and upper faces of the slab $ABCD$ must balance the CORIOLIS force. The friction F_+ on the upper face has the same sign as V_+ because the fluid situated above the slab moves faster than the fluid within the slab and tends to drag it along. In contrast, the friction F_- exerted by the fluid underneath the slab on the lower face has the opposite sign than V_-. For the sum of the forces F_+ and F_- to balance the CORIOLIS force, which is orthogonal to $(V_+ + V_-)/2$, these two forces and the two velocities V_+ and V_- must be oriented in different directions. The forces shown in this figure are shifted to the center of the slab $ABCD$ to highlight their sum. The velocities V_+ and V_-, however, are applied to the center of the upper and lower faces, respectively.

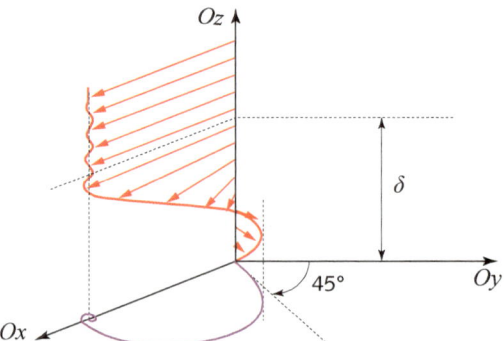

Figure 2.9 - Three-dimensional sketch of the distribution of air velocity across and above the EKMAN layer. At the interface, the velocity within the EKMAN layer must match that of the geostrophic flow. The EKMAN spiral is the projection onto the horizontal (Ox, Oy) plane of this three-dimensional distribution.

Figure 2.9 shows several properties characteristic of the EKMAN layer that we can summarize as follows: the thickness of this layer, which is of the order of $(v_{eff}/\omega)^{1/2}$, where v_{eff} denotes an effective kinematic viscosity due to the turbulent movement within this layer, is relatively modest (several hundred meters), despite the amplification of v_{eff} by small-scale turbulence. In the appendix, we show that v_{eff} can in fact be much greater than the molecular viscosity—by as much as three or four orders of magnitude. Also apparent in figure 2.9 are weak oscillations in the wind above the boundary layer. Finally, note that the wind direction at the ground makes an angle of 45° with respect to the wind at altitudes greater than the thickness of the atmospheric boundary layer.

I2.3 - Concept of boundary layer

Consider a solid body at rest and with characteristic dimension L. This body is placed in a fluid flowing with uniform velocity $\mathbf{V}$. Very close to the solid-fluid boundary, molecular motion very quickly reduces the difference between the velocity of the fluid and that of the solid boundary to zero (the time for this to happen is of the order of $d/c \approx 10^{-9}$ s, where d is the mean free path of molecules and c is their typical speed). Thus, on macroscopic scales, viscous fluids cannot slide without friction over solids. This implies that the macroscopic velocity of a fluid must be the same as that of the solid boundary, which is zero in this case. This constraint represents a strong perturbation with respect to the fluid that is farther from the solid boundary; viscosity resorbs this perturbation by diffusing it within the fluid. More precisely, if we denote the characteristic length of a body by L and the characteristic velocity of the fluid particles by V, the time for the latter to travel the length of the body is of the order of L/V. This is the timescale over which the viscosity must diffuse the zero-velocity perturbation at the solid boundary.

The characteristic time for momentum transport by kinematic viscosity v can be estimated from the classical law governing all diffusion phenomena: the range of momentum transport over time t is of the order of $(vt)^{1/2}$. Denote by δ the thickness of the fluid layer next to the solid boundary (called the boundary layer), where the kinetic viscosity v is capable of transporting momentum. The characteristic time for this transport by viscous diffusion is δ^2/v. The layer within which viscosity can non-negligibly alter the momentum must be such that the two times just mentioned, L/V and δ^2/v, are of the same order of magnitude. We can thus estimate the ratio between the thickness δ of the boundary layer and the length L: $\delta/L \approx (v/VL)^{1/2}$.

The REYNOLDS number, denoted Re, is the name typically given to the dimensionless number VL/v. This number is named in honor of the Irish physicist Osborne REYNOLDS (1842–1912) who was the first to introduce it in 1883. This parameter is of primordial importance in fluid mechanics. From low-viscosity fluids, such as air and water, to length and speed scales that characterize natural flows or those produced in laboratory experiments, this number is almost always very large (10^4 to 10^8), so that the ratio δ/L is often of the order of 10^{-3} or less. In such conditions, the thickness of the boundary layer is very small with respect to the characteristic dimensions of any obstacles that may be inserted into the flow.

I2.4 - The atmospheric boundary layer

Let us apply the estimate presented in insert I2.3 to flow within the atmosphere. Using the actual value for the kinematic viscosity of air (of the order of $10^{-5}\,\mathrm{m^2\,s^{-1}}$), we obtain a thickness for the boundary layer of the order of a few tens of meters. In reality, the boundary layer at the ground is a few hundred meters thick. The difference between the actual thickness and that deduced from the viscosity due to molecular motion is due to the fact that turbulence ensures a much more efficient mixing of the various fluid layers and a much better momentum transfer than does molecular motion itself. The net result is as if the effective viscosity of air were about 10^4 times greater than the molecular viscosity. The kinetic theory of gas (see insert I1.3) estimates this latter quantity to be $v \approx cd$. Let us use u to indicate the speed typical of turbulent fluctuations within the boundary layer. Because instabilities and turbulence appear within the boundary layer, let us also assume that the thickness δ of the boundary layer represents the proper length scale to characterize turbulent motion within the boundary layer. By analogy with the kinetic theory of gas, we may think that the product $u\delta$ provides a good estimate of the effective viscosity due to turbulence. In this case, the estimate of the thickness δ_{eff} of the turbulent boundary layer becomes $\delta_{eff}/L \approx (u\delta_{eff}/VL)^{1/2}$, which leads to $\delta_{eff}/L \approx u/V$.

In practice, the ratio u/V approaches $1/10$ in the distant atmosphere and within the boundary layer. Because L can be nothing but the thickness of the troposphere (about 10 km), this approach leads to a boundary layer several hundred meters thick.

This atmospheric boundary layer, although it is much thicker than the boundary layer of air flowing over the wing of airplanes, remains nevertheless quite thin with respect to the breadth of a low-pressure zone: 500 m vs 500 km. This justifies that, in meteorological models, we distinguish between regions far from the ground, where the fluid may be considered nonviscous and friction negligible, and the atmospheric boundary layer, where any analysis must account for the fact that friction is localized.

The properties of the EKMAN layer imply consequences that are both visible and important. For example, they explain a paradox that is apparently well known by inhabitants of mountainous regions. Immediately after a snowfall, the wind at altitude may seem to blow the fresh snow that has fallen on the summits that extend above the atmospheric boundary layer in a direction opposite to that of the ground-level wind, even within valleys that are sufficiently large to not channel the wind along the direction of the valley. This occurs simply because the observer is located somewhere within the angle formed by these two directions.

Moreover, on maps of low-pressure zones, weather services provide networks of isobaric curves closed upon themselves, as well as the wind intensity at all points above the atmospheric boundary layer; namely, at altitudes greater than a few hundred meters. But sailing fans, who would like to know the winds within a ground-level layer a few tens of meters thick, must jump through two more loops: they must correct the approximate 45° difference in angle for the wind at altitude and also correct for its intensity, which varies strongly within the thin sublayer where their sails operate. Add to this that an EKMAN layer is also present in the water over a layer a couple of tens of meters thick under the surface of the wind-blown water. These two EKMAN layers must be accounted for to understand why icebergs in a calm Arctic Ocean systematically drift to the right of the direction in which they are pushed by the wind.

2.2. Quasi-two-dimensional structure of low-pressure zones

The evolution of low-pressure zones, tracked by meteorologists particularly when they are of sufficient intensity that they might become cyclones, are strongly affected by turbulence on their scale. Every day, meteorological bulletins show that this scale is of the order of hundreds to thousands of kilometers, the size of a typical European country. On this scale, we can expect a quasi-two-dimensional (Q2D) structure to appear, where vertical variations become negligible with respect to variations in the horizontal directions (except within the boundary layer). This organization has significant consequences, because the properties of two-dimensional (2D) turbulence differ profoundly from those of three dimensional (3D) turbulence (see appendix).

Let us pause a moment on this difference and begin by recalling that, in three dimensions, eddies are pulverized into smaller structures, which in turn are pulverized into even-smaller structures, and so forth. Each division occurs on a timescale that is of the same order of magnitude as the time required for the initial eddy to complete a single revolution. The kinetic energy associated with these fluctuations is transported by what we call an *energy cascade*. An energy cascade is said to be *direct* when it goes from the largest scales—those of the largest eddies within which turbulence is created from hydrodynamic instabilities, to scales small enough for the viscosity to dissipate the energy into heat. In fact, this dissipation by viscosity is what prevents the energy cascade from continuing and thus determines the size of the smallest observable eddies. The plumes of smoke coming out of a chimney, or from the crater of a volcano (see fig. A.12 in the appendix), are a good example of 3D turbulence subjected to an energy cascade. With the naked eye, we can see clearly the larger eddies. We can measure their duration and we can guess where the medium-sized eddies form, created by the shearing of the larger ones. But we cannot directly perceive the smallest eddies, those for which the viscosity transforms kinetic energy into heat, because they are less than a millimeter in size. The experimental observation of such eddies, both in laboratories as in nature, has become a classic experiment that requires the acquisition and treatment of very-high-resolution data.

In contrast, in two dimensions, eddies of the same sign mutually induce themselves to take up velocities such that they are swept into rotation around each other in the plane of flow. Convincing yourself of this simply requires observing a single pair of such eddies (see fig. A.6 in the appendix). In each pair thus formed, each eddy wraps around its partner, and the pair rapidly evolves into a new eddy that is twice the size of the original individual eddies, even before the vorticity[7] is homogenized within it. This pairing process is rapid, because its characteristic timescale is nothing more than the period of revolution of the initial eddies, which is much shorter than the timescale of viscous diffusion (the ratio between these times is none other than the Reynolds number, which is much greater than unity for atmospheric turbulence)[8]. If sufficiently close together, the large, newly formed eddies again pair off and the pairing process begins anew. These successive pairings can continue for a long time before viscosity can intervene. As for the kinetic energy of this turbulent 2D motion, we speak of an inverse energy cascade[9] because, in contrast to what happens in 3D turbulence, the energy is channeled toward ever larger scales.

In laboratory experiments on 2D turbulence, the energy reaches the largest scales, those of the basin[10] or of the model under study. It is then dissipated in the boundary layers along the container walls. In the case of the atmosphere, which has no finite horizontal dimension analogous to that of the laboratory basin, the friction along the ground or oceans, relayed by the Ekman layer, dissipates the energy. We shall return to this question to estimate the size attainable by a large Q2D depression (i.e., low-pressure zone) but, first off, between the 2D and 3D regimes, which have several opposing characteristic properties, we must determine the intermediate scale at which the transition occurs between 2D dynamics and 3D dynamics. We shall see that this scale is both significantly larger than the thickness of the troposphere and significantly smaller than the radius of Earth.

Let us begin by examining the circumstances in which atmospheric depressions can find themselves in Q2D conditions. These depressions are shaped like large rotating discs whose thickness (roughly 10 km) is much less than their diameter (500 to 1000 km). This immediately justifies the assumption that vertical displacements are significantly smaller than horizontal displacements, the latter of

---

7   The word *vorticity* refers to the rate of rotation of the vortex, which is a scalar quantity like kinetic energy. Its local value within a mass of fluid rotating with angular velocity ω is $\omega^2/2$. The rotating mass of fluid represents what is typically called an *eddy*, which, during its lifetime, may be considered as a single object.

8   The kinematic viscosity of air is of the order of $10^{-4}\,m^2\,s^{-1}$, so the Reynolds number for a 500-km-wide low-pressure zone where there air velocity is 100 km h^{-1} can be as high as 10^7.

9   This inverse energy cascade is sometime confused with the butterfly effect, which is presented and discussed in the appendix. Strictly speaking, the butterfly effect represents the cascade that proceeds systematically (from the small scales to the large) for any differences between the initial conditions, whether the turbulence be 3D or 2D. It must therefore not be confused with the energy cascade, which is inversed only in 2D dynamics.

10 In a laboratory, where the length scale is of the order of meters, it is practically impossible to implement a 2D or Q2D regime by using air. However, with water, in rapidly rotating basins or canals, we manage to implement and observe these regimes. An example appears in figure A.13 in the appendix.

which benefit from large areas for any eventual accelerations. But beyond this, once the CORIOLIS acceleration becomes dominant, these discs are traversed by waves that propagate in the local vertical direction and prevent the depression from splitting into horizontal sublayers. These waves are called *inertial waves* and are not well known outside of specialists' circles. Their generating mechanism and their properties are summarized in insert I2.5. They have important consequences; notably the tendency to structure large-scale flow into columns, called TAYLOR *columns*. In the case of atmospheric depressions, where the height of the columns (around 10 km) remains much less than their diameter (500 to 1000 km), the use of the word *column* may be surprising. The appellation is justified because inertial waves are responsible for the vertical cohesion across the entire thickness of these rotating structures.

So why then does the CORIOLIS acceleration dominate horizontal motion within large depressions? Imagine an atmospheric depression of large diameter L and moderate speed V, such that the ROSSBY number[11] $Ro = V/\omega L$ is much less than unity, or such that the time L/V for a fluid particle to make a complete revolution is much longer than the characteristic time of the CORIOLIS force ($1/\omega$ where $\omega = \Omega\sin\theta$, with Ω being the angular velocity of Earth and θ being the latitude). Consider a precise example. Place ourselves at medium latitude, where $1/\omega = 50$ h (about two days), because the strength of the CORIOLIS force here is about half that at the poles. Now consider a large, slow depression whose diameter is about 1000 km and whose peripheral speed is about 5 km h^{-1}. The time to cross this depression in the horizontal direction ($L/V \approx 200$ h) is four times longer than the characteristic CORIOLIS acceleration time (50 h). In this example, the CORIOLIS acceleration acts four times faster than the other horizontal accelerations over the entire path of the fluid particles: we can thus say that it is four times stronger. This ratio of about four is not extremely large.[12] Nevertheless, it suffices to justify the pertinence, or even dominance, of the CORIOLIS acceleration and the existence of inertial waves. However, overemphasizing this reasoning to the point of deducing that the other components of acceleration are negligible is not justified. This implies in particular that the pairing process and the inverse energy cascade remain significant. To summarize, it appears that, in sufficiently large atmospheric depressions, inertial waves exist and can combine with other Q2D inertial effects, notably the pairing process and the inverse energy cascade.

---

11 The ROSSBY number $Ro = V/\omega L$ is the ratio between the characteristic timescale of the CORIOLIS force ($\approx 1/\omega$) and the time for a complete revolution of the eddy ($\approx L/V$) or the time necessary for the CORIOLIS acceleration alone to generate a significant change in speed. As we have seen, the period of revolution depends on latitude. In atmospheric depressions, the ROSSBY number can be less than 10^{-3}, which means that the acceleration of the relative motion is negligible compared to the CORIOLIS acceleration. This number is named in honor of the Swedish-American meteorologist Carl-Gustaf Arvid ROSSBY (1898–1957).

12 The preponderance of the CORIOLIS acceleration and the pertinence of the inertial waves are clearer still on other planets larger than Earth. On Jupiter, the Great Red Spot is considered to be a TAYLOR column.

I2.5 - Inertial waves and TAYLOR columns

Consider a horizontal fluid layer rotating as a solid with an angular velocity ω about a local vertical oriented in the direction Oz. This fluid domain could represent part of the troposphere or part of the ocean where the bottom is flat. Now imagine within this rotating fluid a relative motion on a large scale L imposed in horizontal planes and with a modest characteristic speed V so that the ROSSBY number $Ro = V/\omega L$ is much less than unity. More precisely, imagine that this relative motion were a cyclonic rotation with angular velocity $\omega' = V/L \ll \omega$, which satisfies the geostrophic equilibrium because $Ro \ll 1$. The low-pressure zone within this eddy structure due to the CORIOLIS force is of the order of $\rho \omega VL$.

Now suppose that the pressure within this low-pressure zone were not uniformly distributed in the vertical direction and that the angular speed ω' of the cyclonic rotation increases by $\delta\omega'$ over an altitude δz. At the altitude $z + \delta z$, the atmospheric pressure would decrease by $\rho \omega L^2 \delta\omega'$ from that at the altitude z. A pressure difference thus exists between the two horizontal planes, which pushes the more slowly rotating fluid toward the lower pressure (i.e., upward).

During the time δt, a variation in vertical speed δw is generated so that the vertical acceleration balances the upward push due to the pressure difference. This obliges $\delta w/\delta t$ to be of the same order of magnitude as $\omega L^2 \delta\omega'/\delta z$. In so doing, the upward vertical motion transports the weaker angular momentum from the lower layer, thus tending to absorb the difference $\delta\omega'$ in angular velocity and to homogenize the vertical distribution of velocity.

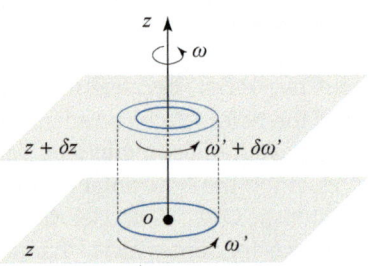

Quantitatively, this vertical transport must be such that the variation in angular speed $\delta\omega'$ over the time δt balances the torque due to the difference between the CORIOLIS forces at the altitudes z and $z + \delta z$. This leads to

$$\frac{\delta\omega'}{\delta t} \approx \omega \frac{\delta w}{\delta z} \approx \omega^2 L^2 \frac{\delta\omega'\delta t}{(\delta z)^2}.$$

The wave travels δz in the time δt and thus has the wave speed, or celerity

$$c \approx \frac{\delta z}{\delta t} \approx \omega L.$$

where the term celerity means the speed of the progressing wave with respect to stationary liquid. As a result, in all fluids that rotate almost as a solid, such as the atmosphere and the oceans, any eventual differences in speed between the various heights tend to be cancelled by inertial waves. Across a layer of thickness h, this smoothing is efficient if the transit time of the wave (h/c) is significantly shorter than the period of cyclonic rotation; in other words, if $h/L \ll 1/Ro$. This condition is clearly satisfied in the troposphere on scales where $Ro \ll 1$.

Another remark is required to complete the reasoning of the preceding paragraphs and that of insert I2.5: in the example chosen, this reasoning suggests that the celerity of inertial waves is of the order of $c \approx \omega L \approx 20$ km h^{-1}. Consequently, the vertical range of these waves during a complete revolution of a depression ($L/V \approx 200$ h) would be of the order of $cL/V \approx \omega L^2/V$, or about 4000 km. This is much greater than the thickness of the troposphere. In other words, these waves travel through the troposphere (thickness $h \approx 10$ km) in about $h/c \approx 0.5$ h. This time is so much shorter than all the other timescales characteristic of the depression (50 h and 200 h) that this crossing can be considered as quasi-instantaneous. The last remark confirms clearly that, for the scales on which inertial waves exist, they ensure vertical cohesion across the entire thickness of the troposphere.

Thinking in terms of images, we can compare inertial waves to vibrations that propagate along virtual rubber bands stretched across the atmosphere between the ground and outer space, a bit like guitar strings. These waves transport energy in the vertical direction and attenuate any eventual differences in air speed between the various altitudes; they thus manage to maintain the proper correlation between the speeds in the various horizontal planes. The pressure is allowed to follow the hydrostatic law in the vertical direction, but it imposes geostrophic equilibrium in the horizontal directions. In practice, these waves are what prevent the large depressions from becoming flaked with planar horizontal sublayers that can slide over each other like discs or like sloppily stacked pancakes (see insert I2.5). On the contrary, by homogenizing the speed along all local vertical directions and by imposing a columnar structure, the inertial waves play a role analogous to that of a axial rod onto which discs are stacked; like this rod, they prevent the various sublayers from sliding over each other.

All significant variations in topography over horizontal distances of the order of hundreds of kilometers generates the same effect and structures the local atmosphere into vertical columns called TAYLOR *columns*.[13] In the appendix, figure A.9 shows an example of such a column over the island of Madeira in the Atlantic Ocean.

Thus, Q2D dynamics is the overriding effect governing the large-scale evolution of atmospheric depressions. This leads to an inverse energy cascade within depressions that can be thought of as discs that are neatly stacked in the vertical direction but nevertheless affected by an internal mixture of 3D turbulence on small scales. Notwithstanding the curvature of Earth, this reasoning could lead one to think that rotating Q2D structures subjected to this inverse cascade would continue to grow until they occupied the entire space available; in this case the entire troposphere. Were this the case, one giant low-pressure zone would develop on one side of Earth with a high-pressure zone on the opposite side. The kinetic energy accumu-

---

13 TAYLOR columns were named after the English fluid-mechanics scientist Sir Geoffrey Ingram TAYLOR (1886–1975), a professor at the University of Cambridge, who discovered them and proposed a theory to explain them.

lated within this single depression would be so great and the winds so violent that this extreme storm would crush everything in its path. Life on Earth would become impossible! Luckily, this situation is not possible.

In fact, a mechanism exists that limits the inverse energy cascade to scales that are significantly smaller than the radius of Earth, and for which we have delayed its examination; it is friction within the EKMAN layer. Its order of magnitude can only be in the neighborhood of the CORIOLIS force and, as we have seen, this force, which is proportional to the angular velocity of Earth projected onto the local vertical direction, is approximately divided by two at the middle latitudes. Its characteristic timescale is thus about 50 h. We can therefore expect that, at the middle latitudes, the inverse energy cascade stops when the horizontal dimension of the atmospheric depression become sufficient for one revolution to take about 50 h. Smaller depressions (but still large enough to have Q2D dynamics) are fast enough to entrain and prolong the inverse energy cascade, but larger depressions, which take longer to execute a complete revolution, can no longer be supplied with energy faster than they are dampened by friction. The time for a complete revolution of the eddy that constitutes the depression (L/V) thus becomes an important magnitude that should be compared with 50 h. A depression whose winds reach 10 km h^{-1} must reach a diameter of about 500 km for its period of revolution to be of the order of 50 h and for friction to stop its further growth. If the wind speed is of the order of 50 km h^{-1} instead of 10 km h^{-1}, the requisite dimension can be multiplied by five to obtain 2500 km. To summarize, friction with the ground or oceans limits the size of depressions to around a thousand kilometers.

2.3. Trajectory and energy of depressions

Atmospheric depressions carry an efficient marker: clouds. These are associated with the condensation of water vapor, which occurs when the local pressure drops below the saturation vapor pressure. We can thus observe depressions with the naked eye, measure their size, and unravel their displacement. At this stage, a remark is warranted to correct a frequent error. Anyone seeing a large mass of clouds moving across a blue sky can be forgiven to think that the wind is responsible for this motion. However, the preceding discussion, based on the geostrophic equilibrium, concludes on the contrary that the wind is actually tangent to the edge of the cloud, because clouds turn about the center of their depression (see fig. 2.7b). It is certainly true that the curving wind associated with the depression is zero at its center and cannot displace the cloud. Thus, other mechanisms must combine to impose this transit through depressions. One of them has already been invoked: the eastward airflow of the jet stream, which is a relatively unstable high-altitude collective-transport phenomenon that globally carries all depression eastward, although they skirt around the more robust high-pressure zones and may undergo intense fluctuations.

Yet another mechanism exists that, under certain conditions, can dominate the mechanism mentioned above. Every depression is equivalent to a gigantic eddy that sweeps into rotation all the surrounding fluid, including that situated outside the depression in question. In particular, each depression drags along neighboring depressions. Thus, all depressions in the troposphere compete to induce within themselves a certain displacement speed. However, the nearer ones clearly are more efficient than those farther away, and in any case the curvature of Earth affects this entrainment. Except for special cases, the result of summing together the numerous contributions[14] implies rather long calculations. However, if we limit ourselves to simple situations to illustrate the speed induced by one depression on another, we can demonstrate this entrainment effect without difficulty. The most immediate example is no doubt that of a pair of 2D eddies of opposite sign situated in the same plane that self-induce a displacement speed along their axis of symmetry (see appendix and fig. A.6).

Finally, we must again invoke another mechanism, linked to the appearance of large-scale spatial variation in pressure within the atmosphere. If, at the middle latitudes, this variation in pressure moves horizontally; for example, like a sort of tide in the sky, the condensation of vapor into droplets follows the region of minimum pressure, whereas the evaporation of the droplets follows the region of maximum pressure. If the mechanism were isolated, the displacement of the cloud edge, which marks the saturation vapor pressure, should be interpreted as the movement of isobaric lines without any accompanying displacement of matter. The cloud speed should thus be interpreted as an indication of the speed of the pressure field and not as an indication of the wind speed at the edge of the cloud.

We often speak of the power of depressions; more precisely, it is their kinetic energy E that should be spoken of. And it is fairly simple to show that their diameter D grows as their energy E increases, and that, for a given energy, depressions are larger in the tropical latitudes than in the temperate regions.[15] The same estimate brings us again to another property already encountered: that wind speed is less in equatorial regions than at the middle latitudes.

Let us take a closer look at the case of tropical cyclones, such as those that develop above the Caribbean Sea or the Bay of Bengal and that are known for their intensity

---

14 The speed induced across a distance by an eddy is expressed by the BIOT-SAVART law, just like the magnetic field induced by an electric current. Together, several eddies thus generate everywhere in space a speed that can be calculated through a summation.

15 In an approximate manner, the diameter D of depressions and their typical speeds V can be related as follows, where Ω denotes the angular velocity of Earth about its axis, θ is the latitude, h is the thickness of the troposphere, and ρ is the density of air. The equality between the characteristic timescales for a complete revolution and friction, itself of the same order of magnitude as the CORIOLIS force, leads to $V/\pi D = \Omega \sin\theta$. The expression for kinetic energy E of the depression is $E = \pi D^2 h\rho V^2/8$. Estimating the diameter and speed is thus elementary: $D = [(8E)/(\pi^3 \rho h \Omega^2 \sin\theta)]^{1/4}$ and $V = [(8\pi \Omega^2 E \sin^2\theta)/(\rho h)]^{1/4}$.

and destructive power, by examining their formation mechanism and internal organization. They appear above the sea when the water temperature reaches or exceeds about 26 °C while a depression passes overhead. The depression is accompanied simultaneously by an ascending movement pumped by natural convection above the warm zone, a convergence of horizontal air currents toward its center, and its own cyclonic rotation. The vertical range of the ascending air, limited by its cooling with altitude, can reach as high as several kilometers. This warm, humid rising air encounters ever colder layers of air, where the water vapor condenses into droplets. This change of state, which is exactly the opposite of evaporation, releases the latent heat[16] of the water vapor, which increases the temperature even more and reinforces the speed of the rising air.

The typical organization of a cyclone is shown schematically in figure 2.10. In the lower part, the curving wind converges toward the center and acquires vertical component to its velocity. Due to the thinning of the air that has captured thermal energy from the sea, this component, which is zero at sea level, becomes larger and larger as the air crosses the EKMAN layer. Beyond the EKMAN layer, the air continues to thin and the vertical speed continues to grow because the air also acquires the latent heat of condensation.

At the center, the eye of the cyclone where the warm air is concentrated is analogous to a chimney whose updraft is linked to the heated air within its vertical conduit. Thus, around the axis of the rather large zone of rotating fluid that constitutes the cyclone, a small sort of central chimney forms. It constitutes a zone of convergence, at the center of which the radial and azimuthal components of the velocity collapse because they must change sign at the axis. Locally, the centrifugal force also collapses so the pressure must increase. This explains both the relative calm of the eye with respect to the intense wind near the wall of this eye and the absence of clouds in the eye of relatively strong cyclones, which makes them easily visible from airplanes flying at higher altitudes (fig. 2.11). Moreover, when the pressure at the center of the eye is not sufficiently strong, which can happen, a descending flow can develop within the axis of the central chimney, in the middle of the ascending flow. When such cyclones pass through an area, the rapid succession of vertical air currents of opposite sign explains why roofs can be violently ripped off and then just as violently deposited nearby their initial position.

---

16 When, at a fixed temperature, water changes state from vapor to liquid, or *vice versa*, a certain amount of energy is released (for condensation) or absorbed (for evaporation). The exchange of heat (called *latent heat*) occurs at constant temperature and differs from the heat exchange associated with variations in an object's temperature (in which case we speak of *sensible heat*). The absorption of the latent heat associated with evaporation explains the relative coolness of forests, or why in the summer we wrap a damp cloth around a glass containing a drink to cool the drink. The latent heat of water near 0 °C is of the order of 334 kJ kg^{-1}.

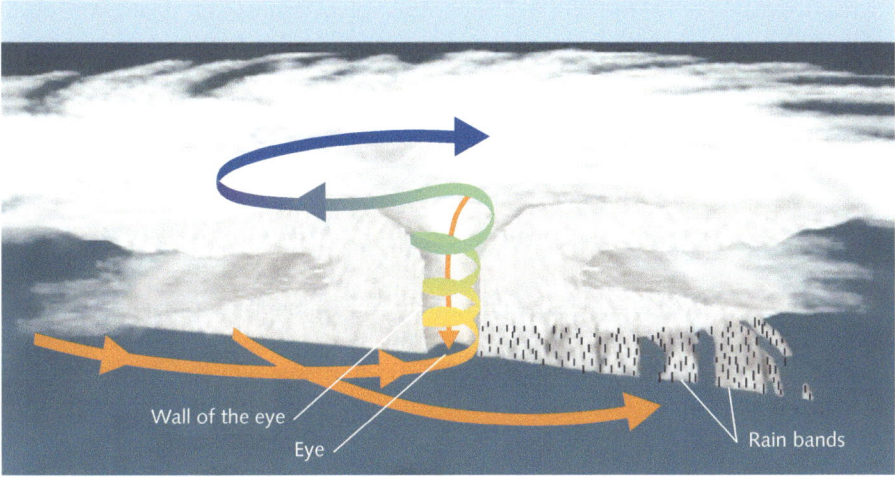

Figure 2.10 - Schematic illustration of the internal organization of a tropical cyclone. The orange arrow in the lower part shows the cyclonic rotation, the yellow-and-green arrow shows the ascending helical movement in the eye, and the blue arrow in the higher part shows the diverging anticyclonic movement. Except within the eye of the cyclone, the speed is essentially contained within horizontal planes, which explains why these structures are considered quasi-two-dimensional (Q2D). [Image from NOAA]

Wall of the eye

Eye

Rain bands

Figure 2.11 - Photograph of the cyclone Floyd off the coast of Florida, taken from a NASA satellite on September 14, 1999 at 12h59. This cyclone caused 57 deaths and damages evaluated at 4.5 billion dollars. [© Hal Pierce/GOES satellite/NOAA]

The important point to retain from this description is the very rapid radial variation in pressure that characterizes the eye of a cyclone, with a minimum at the wall and a relative maximum at the center. The photograph of figure 2.11, taken from above, shows the main characteristics of the cloud ceiling that surrounds the eye: its horizontal breadth and the anticyclonic rotation of the diverging airflow. In the upper reaches of the cyclone, the diverging trajectories of particles moving away from the chimney are deviated to the right by the CORIOLIS force, as indicated in figure 2.10.

3. Periodic phenomena in the atmosphere

To complete this overview of the atmospheric movements, we shall focus now on phenomena that manifest a certain periodicity, starting again with the larger scales, such as those of the seasonal monsoons, and finishing with the diurnal thermals, which are much more localized.

3.1. Monsoons

To illustrate this phenomenon, which is important for the economy of the countries concerned, consider the example of the Indian peninsula (fig. 2.12). The origin of the summer monsoon comes from the fact that the ground warms faster than the seas during the summer and cools faster during winter. In the Northern Hemisphere, the warmed air over the continents in the tropical regions rises and entrains with it the humid air situated above the neighboring oceans. The lifting of the air due to a landscape feature of significant height can thus lead to condensation of this moisture, which results in heavy rains. The inverse occurs in winter and leads to dry monsoons.

Let us look more closely at this. From the beginning of June, the natural convection that develops over the southern point, around the states of Kerala and Tamil Nadu, sucks humid air from the Indian Ocean, which then heads north. A significant fraction of this airflow, situated farther east, passes over the Bay of Bengal, where it becomes saturated with water before arriving at the Ganges estuary and passing over the plain of this sacred river. The foothills of the Himalayas constitute a sort of barrier that deviates the airflow toward the northwest, whereas the rise of the humid air cools it and leads to very heavy rains. This humid air current develops progressively and lasts about two months, from the beginning of June to the end of July. Meghalaya, which is a state situated on the southern slopes of the Himalayas in northeastern India, has the reputation of being among the wettest places on the planet. The countries neighboring India, such as Bangladesh and Burma, are also subjected to this summer monsoon, which strongly influences the cotton, rice, and vegetable-oil agriculture of this immense region. The dates of the beginning and end of this summer monsoon depend strongly on the spring warming and so are always slightly uncertain, although they typically fall within a period of one to two weeks.

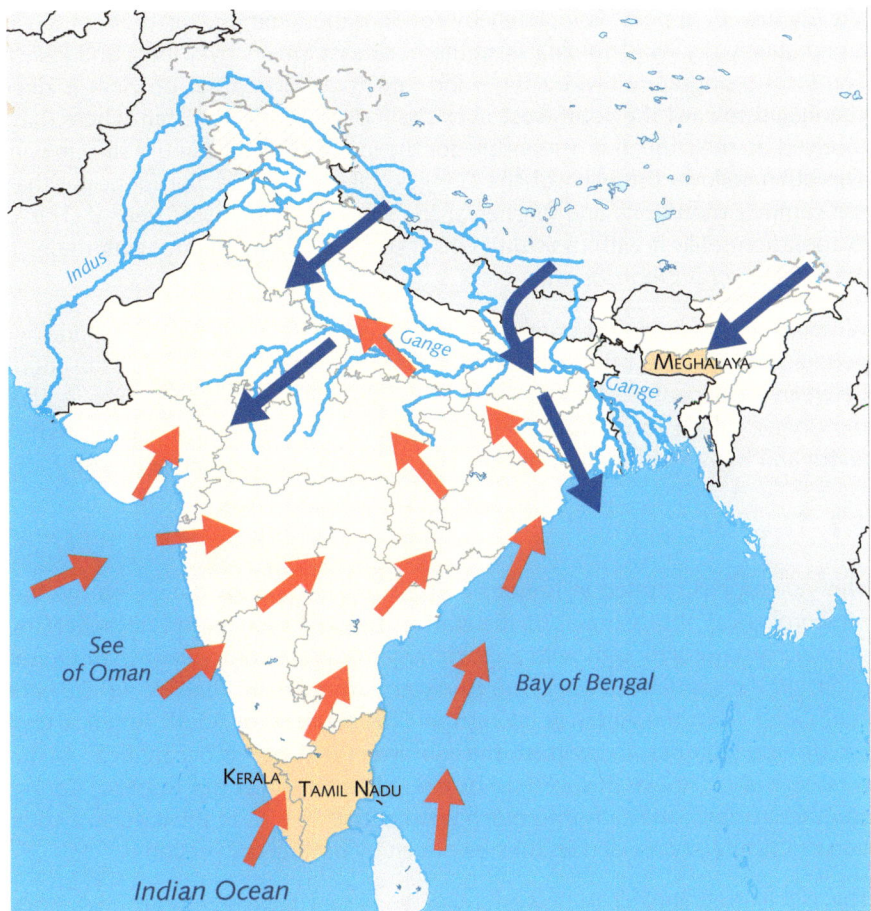

Figure 2.12 - Summer monsoons (red arrows) and winter monsoons (blue arrows) over the Indian peninsula.

At the end of August, with the decrease in daylight and a significant shift toward the equator of the region of maximum sunshine, the thermal engine of the summer monsoon becomes progressively less efficient and the waters of the Indian Ocean and the Bay of Bengal become warmer than the land, which has itself been cooled by the abundant summer rains. Convection can then inverse itself and pump an air current southward. Two other phenomena join with this. One of the two is linked to the presence of the Siberian high-pressure zone on the opposite side of the Himalayas, which reinforces the southward wind and contributes to pushing the air toward the south. The other is the trade winds, which also have a significantly influence in autumn on the southerly displacement of the lower layers of the atmosphere. These winds come from an immense continent and are already relatively dry. As they pass over the Himalayan range, their water content, which is already low, disappears completely by condensation and precipitation. In this way a particularly dry wind coming from the northeast arrives over India and Pakistan. The local topography divides this wind into two flows: One of them is directly channeled toward the southwest over the Indus plain in Pakistan where, having received no moisture, it is responsible for the low water levels in this river in winter. The other follows the plain of the Ganges, taking the direction opposite that of the summer monsoons and finishes at the shores of the Bay of Bengal. The local evaporation refills it with moisture, which supplies the moderate winter rains that we call the *winter monsoons*.

Although the Indian monsoons are often taken as textbook examples, analogous periodic phenomena, often also called *monsoons*, appear elsewhere at the interface between a dry continent and a warm sea. This is the case of the North-American monsoons, which come from interactions between the Gulf of Mexico, which is warm and humid in summer, and the large deserts of the American West.

3.2. The WALKER cell

The WALKER cell, named in honor of Sir Gilbert WALKER who directed the Indian meteorological observatories at the start of the 20th century, is another relatively periodic system, although with a much larger random component than the monsoons. In the southeast Pacific, the HUMBOLDT current (see chap. 5, fig. 5.9) brings cold water from the Antarctic along the Chilean coast of South America toward the tropical latitudes. In contrast, the southwest part of this ocean, near Australia, is rather warm. Above this western ocean at latitudes near the Tropic of Capricorn (southern tropic), the air is thus much warmer than the air to the east, so a convection cell (the WALKER cell) can thus be driven by natural convection.

This cell is itself subject to powerful variations that have significant meteorological consequences. To start with, the seasonal variations in the trade winds, illustrated in figure 2.3, impose on it a rather consistent oscillation whose amplitude is sufficiently large to reinforce or, on the contrary, reduce the westward equatorial airflow at these latitudes. Moreover, the fluctuations of the WALKER cell itself are sufficiently strong that, superimposed on the periodic oscillations, make predictions difficult. Although still subject to debate, their origin seems to be linked to the vagaries of temperature in the waters of the HUMBOLDT current. We are glimpsing here one of the very concrete characteristics of ocean-atmosphere interactions, in this case known under the name *El Niño*, and to which we shall return in more detail in chapter 5 in the paragraph related to ocean currents.

3.3. Thermal winds, Katabatic winds, and anabatic winds

The quasiperiodic phenomena evoked above occur on horizontal length scales of several thousand kilometers, roughly the size of the Indian peninsula or of the

Pacific Ocean. Their period, which is anchored in the rhythm of the seasons, goes hand in hand with this large scale. Over much shorter timescales, such as that of a day, other periodic phenomena develop under the influence of the diurnal variations in temperature. These are the thermal winds, among which are the katabatic winds[17] (this expression comes from the Greek word *katabatikos* (καταβατικός), which means descending the slope) and the anabatic winds (from the Greek anabatikos (ἀναβατικός), which means going up the slope). The simplest manifestation of these winds is known to all. In town as in the countryside, during nice weather, the late-night air is very calm. But as soon as sunlight slightly warms the ground, the RAYLEIGH-BÉNARD instability (see appendix) perturbs the lower layers, the leaves begin to flutter, and the flags begin to flap. This agitation can become sufficiently intense toward midday, and then the evening cools the ground and the lower layers of the atmosphere, restoring the calm and silence of Mother Nature preparing for bed.

In the lowlands, the winds created by this thermal exchange are disordered and turbulent because the characteristic RAYLEIGH numbers (see appendix) are huge, which explains the extreme sensitivity of the wind to the thinning of the air caused by the variations in temperature. In contrast, in mountainous regions, the topography asserts itself by generating katabatic or anabatic winds of sufficient note to merit a comment. During a morning of nice weather, the ground warms faster in the valleys than at the summits or higher plateaus, where the air retains longer its nighttime coolness and elevated density. Shortly after sunrise, the cool heavy air near the summits begins to flow down to the valleys, roughly following the paths of streams and rivers. The katabatic wind is well known by hikers, who encounter it as they cross steep combes: it is present, although feeble, on the sides of the combe, but the flow of cool and therefore relatively heavy air descends primarily along the streambeds and can attain high speeds locally in the axis of the combe. The opposite develops at the beginning of the afternoon, as soon as the solar radiation has warmed the summits, cliffs, or plateaus, even at moderate altitudes. This is the anabatic wind, which is generally weaker than the morning wind because it is spread wider over each side of the streambeds.

Cyclists that ride in mountainous regions are used to these diurnal winds, which are systematically headwinds during morning climbs up to a passes but, on the contrary, are tailwinds during these same climbs in the afternoon. And the inhabitants of certain high valleys, such as the Ubaye valley around Barcelonnette in the southern French Alps, are accustomed during the summer to a rather systematic and constant afternoon wind that blows toward the summits. However, under cer-

---

17 Because the only driving force for wind is sunlight, all winds should be called *thermals*, especially the trade winds and the winds in the WALKER cell, which were described in the preceding paragraphs of this chapter. However, the expression *thermals* has come to mean the diurnal airflow caused by the thinning related to the warming of a mass of air, notably near rocky slopes or cliffs. To be precise and to mark the difference, the terms katabatic and anabatic that are explained in this paragraph should be preferred when the effect of combes is involved.

tain circumstances and notably during a winter high-pressure period, a temperature inversion can occur, which maintains a mass of cold, heavy air in the valley bottoms while moderate winds bring warmer air from the west to the higher altitudes. This situation, which is in contradiction with the average properties of the atmosphere at rest, can be quite stable because the heavier air finds itself underneath the lighter air. Inversions are also well known by hikers, who find themselves dressing lighter at altitude than at the beginning of their hike.

The fog of San Francisco is another of the most well-known consequences of the thermal diurnal winds. The waters of the Pacific near the norther Californian coast, which come from great depths, remain cold all year around, even at the surface. In summer the water temperature rarely exceeds 10 °C, whereas the ambient air can reach and exceed 20 °C in San Francisco and up to 35 °C or more in the agricultural lowlands situated some twenty kilometers farther east. In comparison, at the same latitudes, the waters of the Atlantic Ocean along the European coasts reach up to about 20 °C, which is close to the ambient air temperature. The relatively low temperature at the surface of the Pacific Ocean near San Francisco generates such a significant condensation of the nearby humid air that a cloud almost always covers the ocean near the coast. In fact, this cloud is what hid for so many years the entrance of the San Francisco Bay from sailors coming from Mexico. Apparently, a rather exceptionally sunny day was required in 1775 for the Spanish sailor Juan Manuel DE AYALA to discover the passage that is now known as the Golden Gate and to start exploring this bay.[18]

Almost every summer day, toward the end of morning when the solar radiation has warmed the nearby land, an upward airflow develops several kilometers inland from the coast and sucks in the surrounding air. The cloud sitting on the ocean thus advances toward the coast. It breaches the hills of San Francisco at the lower points and carries on through the streets and between the buildings toward the Bay Bridge, and then onto the bay itself, until this mechanism reverses itself around 3 p.m. In the late afternoon, after the cloud as returned over the ocean, the inhabitants of Berkeley on the far side of the bay watch from the hills one the most famous sunsets of the planet.

Paragliders search the area near south-facing cliffs for thermal winds that can carry them to high into the sky. These upward currents form relatively localized jets since they operate on a horizontal scale comparable to that of a mountain range. They suck in the surrounding air, both that near the summits and from the valleys, and create local low-pressure zones. Let us consider a period of rather nice weather,

---

18 Sir Francis DRAKE reported this bay back in 1579 during his around-the-world voyage, but he did not stop over or explore it, so Juan Manuel DE AYALA is generally credited with its discovery. The discovery of gold along its shores dates from 1848; a nugget considered to be the first discovered is on display at the University of California at Berkeley. The gold rush followed through the American Far West. During this era, the population of San Francisco was only several hundred; now the city is boasts over one million inhabitants.

which implies the presence of a nice high-pressure zone, meaning that the pressure of the atmosphere is relatively high. In the valley, the decrease in pressure due to the thermal wind is insufficient to start the least bit of condensation, so the sky remains clear. In contrast, near the summits where the pressure and temperature are already less than in the valley, this inflow of air is often enough to cause condensation, which leads to the formation of small white cumulus clouds typical of summits. An experienced observer can almost use their appearance, their number, and their intensity as a sort of barometer. A small chain of white cumulus clouds, each centered between the beginning and end of the afternoon on a summit some 1500 to 2500 meters high, is a sure sign of a stable high-pressure zone. On the contrary, cumulus clouds joining together to form a long chain of gray clouds that mass about the summits signifies that the overall pressure is relatively weak and that a local downpour or even a thunderstorm can occur.

The special case of the cloud that encircles the summit of Mont Blanc in France and those of the clouds that encircle all the summits over 4000 m in the Alps is known as one of the signs of a likely forthcoming change in weather. Such cumulus clouds are the result of internal movements modest enough that the cloud ceiling rarely exceeds 3000 m, but yet intense enough that the cloud stretches both over the shady side (also called the *ubac*) and sunny side (also called the *adret*). They can thus exist on the intermediate slopes of the highest mountains but cannot hide their summits. When these clouds appear only on the sunny side, the inhabitants of the valley know that the good weather will continue. However, when they encircle the high summits, the presence on the shady side, which is not subject to thermal winds, is considered a signature of low pressure, which suggests imminent precipitation. If the pressure decreases further, the internal activity within these clouds increases, accompanied by strong ascending airflows; the tops of these clouds can be quite high; enough to hide summits over 4000 m and to cause thunderstorms (see chap. 3).

During good weather, urban airflow also often results from thermal winds. At intersections, certain sides of buildings are sunnier than others depending on the time of day, and the mechanism just described sucks air and creates an air current that does not stop until night. The horizontal direction within the intersection depends on the time of day, because the airflow results both from the quantity of sunlight, and therefore the orientation of the street, from morning to night to the topography of the neighborhood. However, at a given time, the residents can recognize their usual wind. The larger the intersection is and the higher the buildings are, the more significant and perturbing the wind can be. Thus, in the region of Paris, France, the employees of the high-rise financial district *La Défense*, who follow fixed schedules, recognize the wind that sweeps across the plaza with no regard for their elegant hair styles.

3.4. Dappled skies and lee waves

The phenomena described above undergo temporal periodic variations. We can also identify spatial periodic variations within the atmosphere. One of the most well-known examples is that of the dappled sky (fig. 2.13), which is characteristic of pressure distributions marked by a sinusoidal spatial variation around an average value: small cumulus clouds form in the low-pressure areas whereas the sky remains clear and blue near the areas of maximum pressure. More precisely, when the pressure varies sinusoidally in a single direction, this mechanism generates bands of well-aligned clouds; however, when the pressure varies in two perpendicular directions, this mechanism can generate the staggered formation of a dappled sky.

Lee waves are another remarkable effect (fig. 2.14) and occur when the wind, charged with moisture, passes over a relatively short mountain range preceding a long plain. The rise of the cloud on the windward slope is accompanied by a drop in pressure. Together, these two effects create condensation and precipitation; the windward slope of this mountain is well watered and thus covered in lush vegetation. The progressively dried air arrives at the summit, crosses over it, then descends the leeward slope in a gathering wind. In this latter part, it accelerates because the potential energy acquired during the ascension transforms into kinetic energy, just like an oscillating pendulum. As it heads toward the plains, it also warms upon contact with the ambient air, because the air temperature of still air varies by about 6.5 °C per kilometer of elevation (see chap. 1). This warm, dry, and fast-moving air prevents vegetation from taking hold on the leeward slopes, making them systematically more arid than the windward slopes. When the air arrives in the downstream plains, the downward airflow must change its trajectory to horizontal. This imposes a local over-pressure in the plain, which, together with the kinetic energy acquired during the descent, curves the trajectory to give the vertical velocity of this mass of air an upward component. Continuing its journey, which remains horizontal on average, this air starts oscillating between upward and downward motions, exactly analogous to a pendulum, as illustrated in figure 2.14. These oscillations continue and remain visible until friction dissipates a large fraction of the kinetic energy of oscillation.

This phenomenon is well known in North America, especially on the east side of the Rocky Mountains, which border the Great Plains. There, two to three replicas of the initial mountain range are often visible, marked by calm regions between regions with strong winds. In Austria, France, and Switzerland, to the north of the arc of the Alps, this phenomenon generated by a northward wind is often called the *fœhn*. In all these examples, the alternating regions of well-irrigated greenery and windy, arid areas constitute a definite signature of lee waves.

Figure 2.13 - Two variations of dappled skies: In the upper photograph, the sky is covered by altocumulus clouds that form regular bands [© NICHOLAS/Flickr]. In the lower photograph, the clouds form a checkerboard pattern but with a certain irregularity in the distance. [© Emma Jane HOGBIN/Flickr]

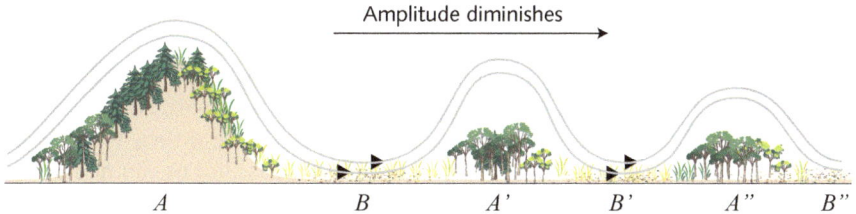

Figure 2.14 - Schematic representation of lee waves. A' and A'' show the replicas of the mountain A. They have no significant wind at ground level and are relatively green in their upwind zones. B' and B'' show the replicas of the valley B; they are almost as windy and arid.

Conclusion

This constantly moving atmosphere is definitely fascinating! To understand its main movements, it has been necessary to distinguish between its diverse scales, from the largest such as the HADLEY and FERREL cells that wrap around the globe, to the more local scales of thermal winds. In each case, energy is furnished by sunlight, which drives the natural convection. The spinning Earth imposes its constraint through the CORIOLIS force, which deviates in the cyclonic direction all movements that occur on a timescale comparable to 24 h. The phenomena are thus selected depending on whether their characteristic timescales are of the same magnitude or less than the length of a day. We have also found that this selection can be transposed to length scales, with the CORIOLIS force becoming dominant for distances exceeding 1000 km, still significant around 500 km, and negligible at less than 100 km. By transforming vertical lines into virtual elastic bands stretched across the atmosphere, the terrestrial rotation generates a family of inertial waves, analogous to the vibrations that propagate along a guitar string. These waves tend to structure the troposphere into columns, at least on scales greater than 500 km, and they impose a quasi-two-dimensional dynamics (Q2D).

Our planet rotates simultaneously about the Sun and about its own axis, which is inclined 23°26' with respect to the axis of its orbit. This imposes a double periodicity on the atmosphere: that of the seasons and that of the day. The effects of each are superimposed on the established and constant circulations on the planet. However, because of the diverse instabilities of atmospheric movements, turbulence, which is always present, also superimposes its will. On small scales, turbulence is three dimensional (3D), at least to a first approximation, and its effect can be understood both as a very efficient mixing mechanism and as an increase in viscosity and friction. In contrast, on the large scales of the atmosphere, where the rotation period is at least of the order of 10 h, motion tends to become quasi-two-dimensional (Q2D). Its original dynamics, which is marked by the energy cascade toward ever-larger scales, play a major role in the formation of large depressions and in the vorticity at the center of cyclones. More locally, monsoons and thermal winds clearly highlight the double rhythm imposed on the terrestrial atmosphere: that of the seasons and that of the day. Spatial periods, such as the lee waves or dappled skies, also appear and contribute to show the high sensitivity of air, this light fluid almost without inertia, to solicitations of all types.

The strongest perturbations, such as devastating cyclones, are tracked, announced, and followed closely by the meteorological services of all countries, with the intent being to warn the population of their dangers. Depressions of medium intensity are both expected, because they bring precipitation that supplies the water tables and the natural freshwater reservoirs, and feared when, if too frequent, they cause flooding. Finally, the softness of light breezes, which seem to flee like the shudders evoked by Charles BAUDELAIRE (see the opening citation of chapter 1), is the stuff of daydreams.

Chapter 3

The vagaries of the atmosphere

The wind is rising!… We must try to live!

(Paul VALÉRY, *The Graveyard by the Sea*)

© Springer International Publishing AG 2017
R. Moreau, *Air and Water*,
DOI 10.1007/978-3-319-65215-3_3

O n horizontal scales significantly smaller than those of the depressions described in chapter 2 (say, less than 100 km), extremely spectacular and often terrifying structures of extreme randomness, such as thunderstorms and tornados, attract a lot of attention. The conditions required for their formation, their internal organization, the mechanisms behind lightning and thunder, and the sometimes devastating consequences of these meteorological phenomena merit some explanation, with the help of the relevant orders of magnitude. Only recently discovered and largely unknown to the public at large, the luminous discharges that accompany thunderstorms all the way to the stratosphere and beyond are also worthy of presentation. The CORIOLIS force, which proves decisive for understanding the large-scale structures studied in the preceding chapter, is no longer significant, but other seeds of rotation exist that are simultaneously strong and localized and that lead to tornados, which we analyze in this chapter. We also focus in this chapter on the large variety of precipitation, eagerly awaited after prolonged dry spells but feared when it brings flooding. Finally, we attempt to discern the limits of weather forecasting and understand the reasons for its uncertainty.

1. Birth and evolution of thunderstorms and tornados

1.1. Dynamics of thunderstorm formation

Before trying to understand thunder and lightning, which are discussed later, let us start by observing a cloud mass susceptible of forming a thunderstorm. Thunderstorms form when a cloud that sucks warm, moisture-laden air up through its base becomes a three-dimensional convection cell and develops strong variations in density in its interior. Most often, this occurs while the cloud is subject to a lift that transports it to an altitude where the temperature is significantly lower. In the lowlands, this lift can come from the cloud's motion toward a warm front where it is lifted over a mass of cool, dense air. More common is its transit over heights within mountainous regions, which leads to an analogous situation. The differences that appear in the density of these structures considerably amplify the convective action, which becomes capable of reaching the high altitudes or low temperatures, thereby facilitating the condensation of water, the formation of large drops, and their descent in localized showers. The thunderstorm appears when, within agitated clouds and between the light and ascending gas phase and the liquid phase made up of much heavier drops pulled down by gravity, friction becomes sufficient to ionize[1] the air, which creates lightning and thunder.

The cloud mass within which convection develops can be an isolated cumulus cloud, such as those that can be seen near summits on a nice afternoon. In these conditions the risk of thunderstorms is very small but, when a depression develops, the cumulus clouds become larger, more numerous, and end up forming a single large group that grows darker and darker as its water-droplet content increases. Within such large cloud masses, which are much more menacing that the small isolated cumulus clouds, is where thunderstorms develop.

Convection in small cumulus clouds (i.e., less than 10 km in size) is initially unicellular, with an ascending airflow (updraft) in the center and a descending airflow (downdraft) at the periphery. However, convection in clouds 10 to 100 km in size tends to be multicellular, with several internal updrafts, although this does not significantly modify the mechanisms governing the cloud's development. As in all depressions, whatever their size, the ascending gaseous phase is progressively dried due to the condensation that accompanies the drop in pressure at high altitudes. With the help of the latent heat given off by the condensation, the cloud cools less than the surrounding air and becomes relatively lighter. Thus, the airflow can attain high speeds and cloud can reach towering altitudes thanks to the added buoyancy force due to its reduced density. This effect tends to lift its mass of water droplets. In contrast, the higher density of the water droplets works in the favor of the downdraft. In fact, the organization of a unicellular convective structure leads to a separation between the central ascending and relatively light column of air and the relatively heavy peripheral regions. The updraft regularly attains altitudes as high as 10 000 m before losing its upward momentum in its fight against gravity and friction. The upper reaches of the cloud mass thus spread out horizontally over large scales as they are carried by the wind (see fig. 3.1).

As liquid droplets circulate within the convection cell and pass through the highest and coldest part of the cloud (recall that the temperature drops by 5 to 6 °C per kilometer in humid air), they grow through condensation and coalescence. They can also freeze to form hailstones. In certain cases, once they have completed several loops around the convection cell, hailstones have endured a succession of partial coalescence in the cloud's periphery during their descent, followed by freezing during their ascent toward the low temperatures of the high troposphere. Some can even become blocks of ice sufficiently large and heavy to cause serious damage

---

1   Ionizing a medium (generally a gas or a liquid), consists of tearing electrons from the molecules or atoms or, on the contrary, of adding extra electrons. These particles, which are initially neutral but now carry an electric charge, are called *ions*. In thunderstorms, this effect can have several causes, such as friction (this is the triboelectric effect) or certain types of electromagnetic radiation. In electrolysis cells, molecules are ionized by the electric field created by applying a voltage across the electrodes.

to airplanes[2] that may be crossing the upper reaches of this nebulous mass. In contrast, during their descent and their superficial melting, part of their liquid mass is ripped off by friction. All these liquid or solid objects that are rapidly circulating around the cloud are thus much smaller and lighter in the lower part of the cloud than in the upper. On the periphery of the cloud, the large drops and hailstones cannot be carried back up, because they are too far from the updraft, which is localized in the center of the cloud. Thus, they drop, pulling some of the nearby air downward as they fall. This precipitation of large water drops and hailstones thus pulls downward toward the ground the air in the periphery of the thunderstorm, thereby forming one of the driving forces behind the peripheral downdraft that complements the densification of the air as it comes in contact with the colder exterior air. This is the mature phase of a thunderstorm (fig. 3.1).

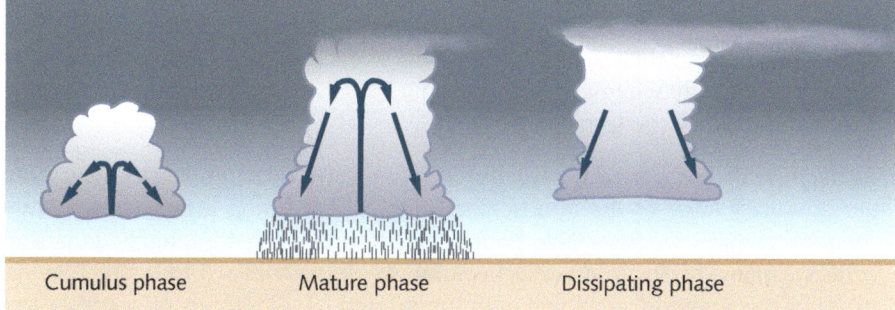

| Cumulus phase | Mature phase | Dissipating phase |

Figure 3.1 - Schematic illustration of the three main phases of the development of a thunderstorm. During the first phase, which consists of a cumulus cloud, note the organization of convection cell with an updraft in the center and a downdraft at the periphery. As the cloud reaches maturity in the second phase, the rain or hail appears preferentially at the periphery. At this point the convection cell reaches extremely high altitudes. Finally, in the third phase, the updraft essentially disappears and the cloud takes on the anvil shape that characterizes the end thunderclouds.

The evaporation of the rain on the ground progressively cools the ground and the nearby layers of air, thus reducing the cause of the thunderstorm; namely, the presence of warm, humid air in a colder environment. Unless warm and humid air arrives from elsewhere to compensate for this local cooling, the end of the thunderstorm is in sight. Progressively, the supply of water and hail to the upper part of the cloud decreases and it unravels. Recognizing the anvil shape that characterizes these dying clouds is rather easy, even from the ground far away. The lifetime of a large, relatively isolated cumulus cloud is often less than an hour. However, when they come from vast multicellular convection structures within large depressions capable of renewing them, these phenomena can follow one another over several hours in an apparently ceaseless manner.

2   The October 5, 2009 issue of *Aviation* shows photographs of the cockpit and wings of a Boeing 737 that was hammered by such blocks of ice after taking off from the Geneva airport in 2003.

1.2. Tornado formation

When a heavy cloud mass acquires sufficient rotation, the convection cell described in the preceding section can develop into a tornado, whose structure is character- ized by an almost-vertical eddy tube extending from the cloud to the ground or the sea (see fig. 3.4), without this affecting significantly its thunderstorm charac- teristics. Rather particular conditions must be met for tornados to form because the rotation cannot be generated by the CORIOLIS force on the scale of such a cell (sev- eral kilometers or less). Moreover, extremely rare conditions are required for a sort of finger that is heavier than the surrounding air to form in the lower part of the storm cell and descend to the ground, concentrating very locally both the rotation drained from the surrounding cloud and the water-droplet content. To obtain these conditions and create a characteristic tornado, several mechanisms must simulta- neously come together. Tornadoes can look rather different and their intensity can vary depending on the relative contribution of each of these mechanisms. At the risk of being overly schematic, we shall limit ourselves in this section to describing one of the most classic tornadoes.

To begin with, between the layers of air within the cloud itself and the ground, the boundary layer of the atmosphere imposes a shearing action, as we saw in chapter 2, which is indissociable from a certain vorticity. To thoroughly understand vorticity, mentally isolate a vertical column of air, such as AB in figure 3.2a. The speed of its extremity in contact with the ground (A) is zero, its opposite extrem- ity (B) moves at the speed of the wind at that altitude. To move from position AB to position AB', this column must rotate about the axis perpendicular to the page and passing through point A. This shows that a certain vortex intensity is present everywhere within this sheared boundary layer. However, in this large-REYNOLDS- number flow (the REYNOLDS number for atmospheric flow [3] was estimated in the preceding chapter to be around 10^7), the vorticity cannot remain uniformly distrib- uted because inertial effects concentrate it (see the example in the appendix of the KELVIN-HELMHOLTZ instability) by forming clearly separated horizontal roll vortices that inevitably align themselves in the direction of the external wind (fig. 3.2b). This evolution toward parallel rolls aligned in the direction of the external wind velocity is characteristic of the boundary layers and is one of the known steps in the transition toward turbulence of this type of airflow.

In addition, we saw in the preceding section that thunderstorms are accompanied by at least one intense updraft, which drains the surrounding air by forcing it to converge at center of the thunderstorm. The rolls generated by horizontal shear forces in the boundary layer can thus be pulled upward by the updraft, changing

---

3   The interpretation of the REYNOLDS number given in chapter 2 is as follows: it measures the ratio between the characteristic time of viscous friction and the time required for air particles to com- plete one revolution in a typical eddy. This is strictly equivalent to saying that it measures the ratio of inertial forces ($m\gamma$ in the equation of motion) to viscous frictional forces.

their horizontal axis of rotation to vertical (fig. 3.2c) and carrying toward the heights all the material that surrounds them (dust and vegetation). This mechanism, which initiates a vertical rotation in the updraft of a thunderstorm cell by transferring to its vertical axis the angular momentum of the horizontal roll vortices of the boundary layer, is analogous to the mechanism that generates vortices in sinks and bathtubs, where the vorticity of the boundary layer on the curved surface of the sink is gathered around the axis of the drain. Note that, because this mechanism is found everywhere a thunderstorm forms, it alone cannot explain the formation of tornados. However, it nonetheless constitutes one of the potential sources of rotation about a vertical axis.

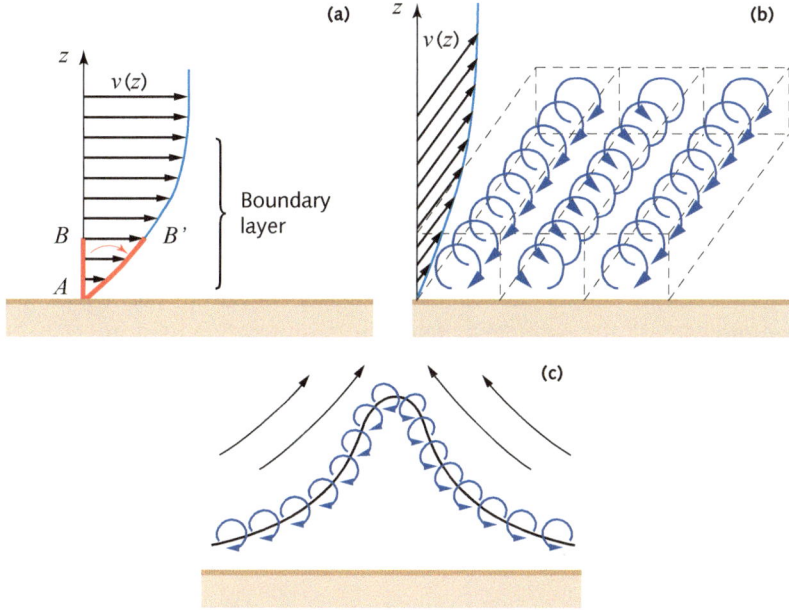

Figure 3.2 - A mechanism in the boundary layer of the atmosphere below a thunderstorm cell leading to the formation of eddies with a vertical axis: **(a)** the presence of horizontal vorticity associated with the shear forces that make AB turn into AB'; **(b)** mature stage of instabilities in the boundary layer, leading to a network of horizontal rolls aligned along the direction of the wind velocity; **(c)** lifting of one of the horizontal rolls by the updraft, leading to an eddy with a vertical component to its axis.

Another source for creating vertical eddies that is more frequent in some regions of the globe than in others is the shearing between two horizontal airflows moving in opposite directions: a warm, humid wind and a cold dry wind (fig. 3.3) circulating horizontally side by side, but in opposite directions. In the United States, the states of Florida, Louisiana, Texas, Oklahoma, Kansas, and Nebraska constitute a sort of prototype of this type of region where such shearing is common and where, for this reason, tornados often form. This effect has earned this region the name of Tornado Alley.[4] The warm, humid wind comes from the Gulf of Mexico to the south; the cold wind comes from Canada to the north after accelerating through the horizontal funnel between the Rocky Mountains and the Appalachian Mountains. When these winds are well established, the layer of air that separates them is sheared, leading to vorticity with a vertical axis. As with all large-REYNOLDS-number shear structures, this vorticity does not remain homogeneously distributed along the surface of maximum shear between these winds (see plane MN in fig. 3.3). On the contrary, the KELVIN–HELMHOLTZ instability mechanism described in the appendix (see fig. A.3 and A.4) concentrates it in eddies with vertical axes, which serve to prime tornados.

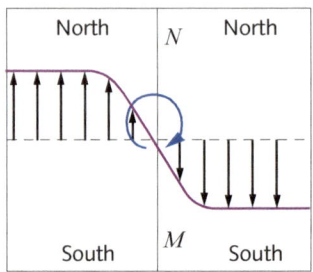

Figure 3.3 - Presence of shearing in a horizontal plane at the interface between horizontal winds moving in opposite directions. Due to the KELVIN-HELMHOLTZ instability, the vorticity associated with the shearing is concentrated at several points where it creates vertical eddies of the same sign, which may serve to prime tornados.

The difference in temperature and humidity between the two winds described above provides all the ingredients favoring the formation of storm cells. A lifting of a mass of warm air above cold air suffices to create the conditions conducive to this development. The storm cells thus formed are usually large and multicellular. Their updrafts are already blessed with vertical-axis vorticity that they carry to high altitudes, thereby stretching the lines of eddies, which has the effect of increasing the local angular velocity.[5] But, in addition, these rotating storm structures gather even more vorticity by sweeping up in their updrafts the vorticity created by the horizontal shearing of the boundary layer of the atmosphere.

All that is left to understand is how the lower part of the cloud can form a sort of downward-pointing rotating finger situated below the updraft, and how this finger can become a continuous rotating columnar structure that reaches all the way to the ground. Liken the lower boundary of the cloud to an interface between the internal medium of the cloud, which carries opaque liquid droplets, and the atmosphere situated below, which is purely gaseous and transparent. Almost everywhere, this interface is stable and remains horizontal. This implies that, despite its cargo of droplets, the cloud matter is globally lighter than the matter below it. Explaining the formation of the descending finger based on the RAYLEIGH-TAYLOR

4　Approximately 1000 tornadoes occur each year in the United States, of which over a third are triggered in Tornado Alley.

5　Any small domain about an eddy line possesses the following important properties: when it elongates its cross section diminishes because mass must be conserved, and its angular velocity increases to conserve its angular momentum.

instability described in the appendix would thus be erroneous. Furthermore, this descending finger appears precisely in the region where the updraft drives the convection cell, which would naturally oppose the eventual descent of this fluid domain. How then can we explain that this rotating finger forms, elongates rapidly downward, and finally reaches the ground?

The explanation comes from the rotation of the updraft within the center of the storm cell, which is in certain respects analogous to the rotation of the eye of a cyclone. This intense rotation generates a local centripetal depression, which is added to the depression already present on the larger scale within the cloud mass. Where the pressure is lowest, namely, on the axis of the rotating updraft centered in the cloud, condensation is locally much greater than in nearby regions. The drops within this region near the axis can thus grow until, in spite of the updraft, they become large enough to fall, taking with them their surrounding gas. A descending jet colored darkly by its content of large drops thus emerges from the middle of the updraft, but only within its central and lowest part. The added depression that can provoke this plunge is, in fact, sufficient only within a central portion of the rotating updraft, very near its axis. It is under these circumstances (illustrated schematically in fig. 3.4), which are indeed quite exceptional, that a rotating descending finger can appear and stretch all the way to the ground.

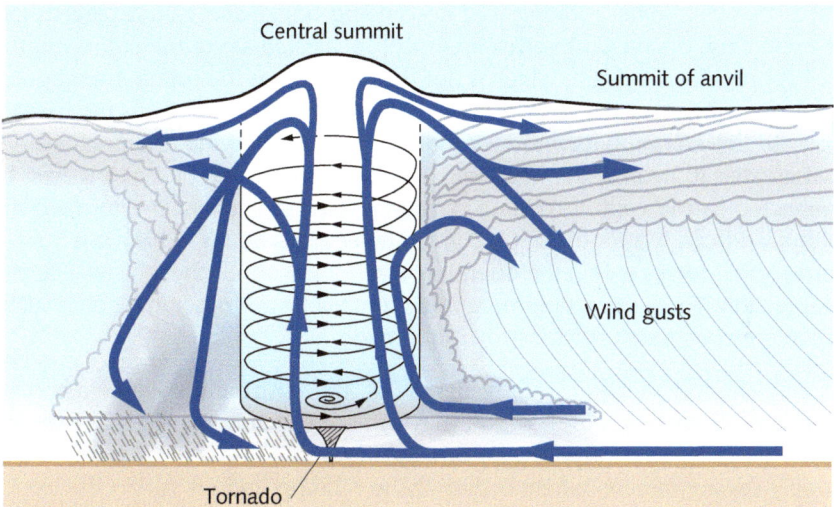

Figure 3.4 - Internal organization of a storm cell connected to the ground by a tornado. Even if the scales are not rigorously respected, note the small size of the tornado (gray hashed area) with respect to the rotating updraft (light blue color). The height of the tornado can be on the order of 10 to 100 m, the total thickness of the cloud mass can be greater than 5000 m. The diameter of the tornado can be on the order of 2 to 10 m, whereas that of the updraft can attain 50 to 100 m.

The descending, rotating finger develops rapidly and resembles a rapidly vertical rotating funnel. It is neither stable nor axisymmetric because its low point is carried by the swirling winds in the base of the cloud and, as we shall see, its point rapidly reaches the ground. The vortex tube is rather large at the top but, at its lowest part, the airflow converging toward the axis reduces the vortex cross section to zero at its lowest point. To maintain a constant angular momentum (ωr^2), the angular velocity ω increases in the lower part of the vortex and the centripetal depression increases more and more. Above the lowest point of the finger, the pressure is minimal and differs the most from the saturation vapor pressure. The condensation thus increases ever more in this zone, charging this zone with more and more water drops, making it even heavier and darker. In contrast, outside the vortex, the pressure is higher, water remains in the vapor phase, and the air remains transparent. Gravity thus rapidly pulls the heavy finger down until it touches the ground.

The precursor stage to the rotating and descending finger is short and quickly results in a well-established tornado. Once the ground or sea is reached, a very visible vortex tube forms (fig. 3.5), which meteorologists call a *funnel cloud*. It is relatively homogeneous along its height, even if, in the case of longer funnel clouds, a conical shape remains visible, often accompanied by significant distortions. This rotating structure, which has reached its limiting length, proves to be much more stable than the initial rotating and descending finger and can last much longer. The condition of constant vortex intensity[6] is satisfied along the entire length of a well-formed funnel cloud, which is not the case before the rotating finger reaches the ground. This small-diameter, quasi-vertical vortex that connects the base of the cloud to the ground becomes ever more independent of the low-pressure system that created it, even if the latter supplies it with water droplets. However, being connected to the central rotating column of the depression, the tornado moves with it and can trace out rather curious trajectories during which it is subject to strong shear forces within the boundary layer of the atmosphere, as well as significant torsion. These strong perturbations, combined with other influences such as a rise in pressure, can finally destroy the tornado.

As we said and as suggested by figure 3.4, the horizontal scale of a tornado is much smaller than that of the updraft, which itself occupies only a small area at the center of the entire storm cloud. The diameter of the tornado can be on the order of only several meters, whereas that of the rotating updraft within the cloud can get to be a hundred meters or so, all contained within a cloud structure covering several kilometers. A structure as unique as an updraft within a cloud is difficult to detect in a dark cloudy sky and so is feared by airplane pilots who might cross

---

6   Like flow rate in a pipe, the magnitude of electric current in a conductor, or magnetic flux, vortex intensity cannot vary along the length of a vortex tube. Thus, a vortex tube is constrained to assume particular forms, either by looping back on itself like a smoke ring, by going from one wall to the other, or (in the case of a tornado) by going from the interior of the cloud all the way until it makes contact with the ground or the sea. This property is shared by all fluxes of zero-divergence vector fields. These fields are called *conservative*.

it. If they do so, they experience alternating plunges and rapid climbs. Until the airplane exits the depression and regains a stable lift, the crew and passengers can be subjected to a jarring ride.

Figure 3.5 - Tornado funnel cloud and the waterspout that it creates. Photograph taken on 30/01/03 in the Mediterranean off the coast of Calvi. [© Météo France/Michel Luciani]

Whereas the time required for the funnel cloud to form is on the order of several minutes, the duration of a tornado can be on the order of several hours, during which it can cover several kilometers, destroying along its way light buildings and wreaking havoc on vegetation. The tornado shown in figure 3.5 is not particularly big, because its diameter is on the order of several meters, nor is it particularly powerful. Its funnel is rather long; around several hundred meters. The most devastating tornados can be recognized by their short, fat funnels (tens of meters long by tens of meters wide). Estimates by meteorologists imply that the tangential velocity at the edge of a large funnel can exceed 300 km h^{-1}, or even 400. Given these conditions, Bernoulli's law allows us to estimate that the internal pressure is reduced by 8% to 10% with respect to the external pressure, which means it can become less than 920 hPa. The force exerted on a roof with a surface area in the neighborhood of 100 m^2 thus exceeds 10^6 N, which is equivalent to a weight of several tons. We can thus understand how passing tornadoes can rip off roofs and uproot trees. At sea, tornados generally provoke waterspouts by lifting a crown of water[7] several meters into the air, as well as sea animals and fishing boats or

---

7   Normal atmospheric pressure (about 10^5 Pa) is the equivalent of a 10 m column of water. A local perturbation of 10% can thus locally raise water one meter above the sea. This perturbation is proportional to speed square and so becomes very strong at high speeds.

pleasure craft that may be in the vicinity, which is not surprising in view of the photograph of figure 3.5.

The risk of a tornado occurring, as for that of a thunderstorm, can be predicted and announced because the circumstances favoring their formation are known and are relatively specific. But the exact position and the instant that the tornado will form at the base of the cloud, as well as the trajectory that mature tornado will take and the duration between its formation and decay is determined by a host of complex influences working on the medium scales, conferring on it an extremely random character and making it particularly difficult to predict.

2. Sonic and luminal signatures of thunderstorms

In everyday language we often use the words *thunder* and *lightning* to describe overall the manifestations of thunderstorms that are audible and visible from the ground; these words immediately bring to mind the potential for disastrous consequences, such as wildfires and destruction. However, the light flashes emitted by thunderstorms in the high atmosphere, poetically named *sprites*, is much less known and has only recently gained the attention of methodic observation. The physics of these sonic and luminous phenomena is complex and still poses many unresolved questions, particularly concerning the effects of ionizing radiation. We shall thus limit this section to their observation, accompanied by several comments and explanations on the best-known mechanisms that lie behind them.

2.1. Lightning

Let us come back to our examination of cloud structures susceptible to create lightning. These are excited by an intense convective circulation, where the air speed can exceed 100 km h^{-1}. In such clouds, just as in the depressions described in the previous chapter, the thinning of the gas phase and its progressive drying due to condensation locally amplifies the upward vertical forces, which accelerate the updrafts. Very large shear forces appear between these updrafts and the downpour of water drops and hailstones. On the scale of small, relatively heavy objects, this friction ionize the air via the triboelectric effect,[8] which locally creates sparks analogous to the easily visible variety emitted by dry woolen articles that we shake in the dark and that make small crackling sounds. In the daytime, it seems well established that solar radiation also contributes significantly to this phenomenon, but the difficulties encountered by specialists in ionizing radiation have thus

---

8 The triboelectric effect refers to the phenomenon of electrically charging a material by rubbing its surface with another material. A part of the work done by friction is converted into electric energy, which depends strongly on the electrical capacitance of the material. When the threshold corresponding to this capacitance is exceeded, electrons are ripped off to neutralize the positive charges on the rubbing material.

far prevented them from giving a simple explanation for these phenomena or from modeling them with precision.

Imagine a water drop progressively elongated due to friction with the gaseous medium surrounding it. This gas tends to sweep the water drop along in its upward motion, but is not completely successful because of the weight of the water drop, which has become rather large (fig. 3.6a). Because the water drop cannot keep pace with the ascending air, the former exerts downward friction forces on the latter. When these are sufficiently strong to do battle with the electrostatic forces internal to the atoms that make up the gaseous molecules, they become capable of pulling electrons downward and of ionizing certain atoms.[9] These electrons gather in the lowest part of the interface between the air and the liquid drop, causing a negative electric charge to appear. And because the overall electric charge of the drop must remain zero, positive ions (i.e., atoms deprived of some of their electrons) gather around the upper part of the drop. In this way an electric polarization can appear on the scale of the isolated electric drop, as schematically illustrated in figure 3.6a.

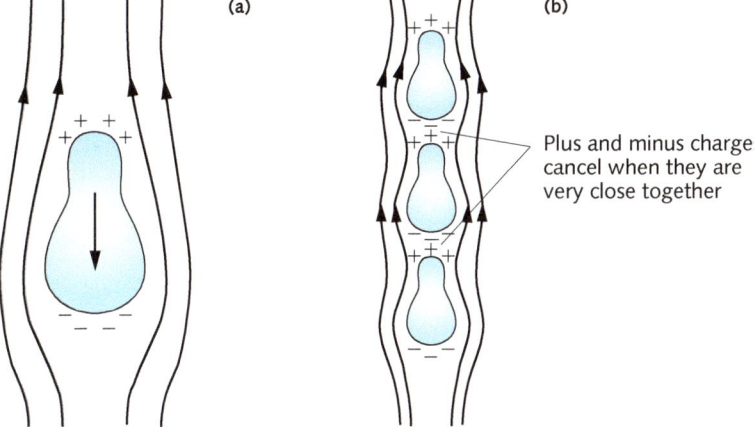

Plus and minus charges cancel when they are very close together

Figure 3.6 - Schematic representation of the mechanisms by which air is ionized by triboelectricity. **(a)** Polarization appears on the scale of a single water drop heavy enough to not keep pace with the ascending air surrounding it. The drop rubs against surrounding air and concentrates electrons around its lower air-water interface. By conservation of total electric change, positive ions concentrate in contrast at the upper air-water interface. **(b)** On the scale of a column of polarized drops, the positive and negative charges between two nearby drops can neutralize each other, thus leading to a large-scale polarization. The shape of water drops on this figure is purely qualitative and does not represent their complex behaviour under influence of friction and capillarity.

---

9   Electrons are relatively easy to remove from atoms because they orbit about the nucleus in multiple layers called *shells* at distances from the center (on the order of 1 Å) that are much greater than the size of the nucleus itself. The same does not hold for the positive protons, which are solidly fixed within the nucleus by the strong nuclear force. The angstrom (Å) is a unit of measure appropriate for atomic scales, and is the order of magnitude typical of an atomic radius. 1 Å is 10^{-1} nanometers.

Even if, most often, the negative-charge density (the electrons) is in fact greater in the lower part of the drop and the positive-charge density is greater in the upper part, the inverse can also occur, although more rarely. In what follows, to simplify the description of this complex phenomenon, we shall continue to consider that the ionization leads to a majority of the negative change accumulating toward the bottom, with the positive charge accumulating toward the top.

The phenomenon thus described concerns not only a single drop but a great many. We can therefore imagine columns of drops within the updraft, all aligned by friction (fig. 3.6b) and all carrying a positive charge on the top and a negative charge on the bottom. In certain places, notably when the length of the drop becomes significantly greater than the distance to its nearest neighbor, the electrons of an upper drop can escape and neutralize the positive ions at the top of the lower drop. In this way, as the phenomenon repeats all along the column, the ionized domain elongates more and more in the vertical direction, with the positive charges concentrating toward the top and the negative charges toward the bottom.

This ionization, which is primed at the scale of a water drop, can become a thoroughly collective phenomenon. The drops that travel upward toward the top of the cloud gather a positive charge near the top. Reciprocally, the drops that descend gather a negative charge in the lower part of the cloud. This latter part can then depolarize itself on large variety of scales, from that of a vertically stretched drop to that of the thickness of the entire cloud. The electric energy thus accumulated can become quite considerable, even though it represents just a fraction of the work done by friction, which itself is only a fraction of the kinetic energy of the convective movement within the cloud.

As electrons are ripped from certain atoms and taken up by positive ions, the atoms change energy levels and thus emit photons, creating sparks that prime the lightning strikes. These sparks produced by weak and localized electrical discharges on the scale of water drops are not yet visible from the ground; this stage is just the precursor to the lightning.

If this ionization process continues, further increasing the electric energy, the electrostatic forces exerted by the elementary charges on each other are transmitted to their carriers; in other words, to the dust, to all the ionized particles, all the way to the largest of the water drops. By repelling each other, the charges of the same sign also repel the matter that carriers them, thus creating within the cloud ionized channels that are almost empty. The only antagonist force capable of limiting the expansion of these channels is the compression of the surroundings, which is necessarily in equilibrium at the outset. This phenomenon of expansion begins extremely fast, but it also lasts only a very short time because the compression of the fluid surrounding the expanding channels limits the expansion and then recontracts the channels. We can compare these very rapid alternating expansions and contractions to explosions followed by implosions. This shows how ionization that starts on very small scales can lead to the formation of extremely long ionized channels, which may be branched and are extremely conducting.[10] These channels transfer more and more negative charges at the lower reaches of the cloud and emit more and more photons (i.e., light), which is when lightning appears.

At this point, the negative electric charge in the lower part of the cloud may create a local electric field capable of separating positive from negative charges in the ground. The positive charges are attracted toward the highest points, such as the tips of church steeples, chimneys, treetops, or the more slender mountain summits. Electrostatics has given us a thorough understanding of the effect of points, which results from a concentration of equipotential lines where the radius of curvature is minimal, leading to a maximal electric field (fig. 3.7).

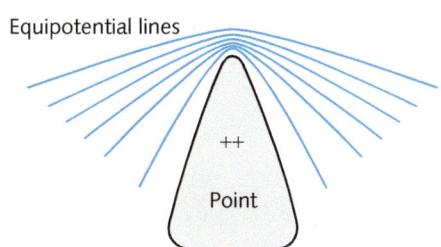

Equipotential lines

Figure 3.7 - Illustration of the point effect, which is due to the squeezing of equipotential lines in the void near the extremity of a positively charged object with a very small radius of curvature.

This effect becomes all the more important as the radius of curvature becomes smaller. In the limiting case, when the radius of curvature goes to zero, such as for a specially sharpened point, the local electric field would go to infinity. Because it is proportional to the electric field, the attractive force exerted by positive ions in points can become locally sufficient to rip electrons from the base of the cloud, thus neutralizing the positive ions in the point. This electric discharge, which occurs through small fluid channels between the cloud and the points, is often called *electrical breakdown* (or simply breakdown). This phenomenon is accompanied by a strong emission of photons, resulting in bright lightning, such as that shown in figure 3.8.

The phenomenon of electrical breakdown is also encountered in other situations; for example, in capacitors, whose electrodes are separated by a strong insulator. Breakdown occurs when the positive charges in the anode and the negative charges in the cathode manage to locally ionize the insulator and neutralize each other. Even if, by construction, the capacitor electrodes are plane or cylindrical, the residual roughness of their surfaces presents very small points where the electric field becomes very strong. It is from these points that the lightning emanates, which

10 The matter within these channels is ionized; this state of matter is called a *plasma*, and its dynamics and transport properties are very complex. A plasma can only exist in extreme conditions, either at high temperatures such as in electric-arc furnaces or in stars, or at very low pressures such as in certain fluorescent tubes. The interested reader can consult the book *Physics of Collisional Plasmas. Introduction to High-Frequency Discharges*, by M. MOISAN and J. PELLETIER, Springer (selected by Grenoble Sciences), 2012.

leads to breakdown. This phenomenon is the subject of intense studies because it constitutes one of the causes of degradation in capacitor efficiency. Another example of electrical breakdown is the electric discharge that an arc welder creates and maintains between his electrodes, which are always of small dimensions to benefit from the point effect to localize the arc, and the metallic pieces being welded, which are typically of much larger dimensions. Electrical breakdown is actively sought by the welder, who controls it by constantly adjusting by hand the distance between the electrode and the pieces being welded so as to keep the arc stable.

Figure 3.8 - Examples of lightning influenced by the point effect: electrical discharges in the cloud above the EIFFEL Tower. Note the presence of other, less-luminous lightning bolts, one of which comes from the top of a chimney. [© M.G. LOPPÉ, June 3,1902/NOAA]

Now that we have become familiar with the point effect, we can understand why the lightning rods consist of relatively sharp metallic objects placed on high buildings, such as steeples. They are connected to grounding electrodes that are buried deep in the ground, preferably in a moist soil so as to ensure a large capacity to dissipate electrical discharges. Figure 3.8 shows very luminous lightning bolts that converge onto the point of the EIFFEL Tower, but it also shows that other, less-luminous lightning bolts can follow strange paths within the cloud.

To complete this simplified description of the phenomena that generate lightning, let us go up by several orders of magnitude. A lightning bolt of average power can occupy a very mobile and fleeting channel of several millimeters in diameter. The length of a lightning bolt can reach up to a kilometer, with the potential difference between its extremities largely exceeding 10^8 V (100 million volts), and the electric current during the discharge easily surpassing 10^4 A (10000 amperes). These values impose large cross sections for the metallic cables used with lightning rods. The instantaneous power of lightning can thus exceed 10^{12} W (1 million megawatts), which is enormous. However, it lasts an extremely short time; much less than a microsecond. Nonetheless, it suffices to raise the temperature of the very-low-density matter within the channel to 30000 K, which is compatible with the emission of such an intense light.

2.2. Thunder

The thunderclap represents only a very small part of the energy of a thunderstorm system: the fraction situated within the audible spectrum of the energy of the shockwaves generated by the electrical breakdowns. These shockwaves are produced by the extremely fast dilations of the ionized channel, which we compared to explosions followed by violent implosions, each emitting sound waves. The sharp, intense, brief sound is due to the superposition during a short time of the sound-waves emitted beforehand over a relatively long period during the formation of the ionized channel. In fact, the ionized channel dilates at a supersonic speed, so that the waves emitted at a given time catch up with those emitted slightly earlier. Their superposition in a thin region and during a short time is what makes the shock.

We shall have the occasion to reexamine the mechanism of the superposition of soundwaves in the next chapter to understand the sonic boom of supersonic airplanes. However, to illustrate it right away, imagine a stagecoach driver cracking a whip in calm air. The sharp crack of the whip is produced only if the driver uses the speed of his arm and a flick of his wrist to make the tip of the whip move at a supersonic speed, at least for a short period. The air surrounding the tip of the whip catches up with the waves already generated, and it is the superposition, at least for a short time, of the soundwaves emitted beforehand over a longer period of time that concentrates their energy and creates the sharp crack of the whip. Without the final flick of the wrist of the driver, the sound of the whip may be audible, but it is reduced to a more continuous whoosh, duller yet longer lasting.

The breakdown of a relatively straight lightning bolt corresponds to a brief and violent thunderclap. In contrast, that of a system of long, branched lightning strikes can be recognized by the rolling sound spread over a certain duration, because the explosions and implosions do not occur at the same time all along the branched system. We know the exercise parents propose to distract their children frightened of thunder: given that the perception of the lightning is quasi-instantaneous, and knowing that sound propagates at speed of approximately 340 m s^{-1}, how far away did the thunderstorm just erupt?

2.3. The lights of the upper atmosphere

The existence in the stratosphere and mesosphere of luminous phenomena associated with thunderstorms in the lower atmosphere was envisioned already in the 1920s, but their methodical observation only really started in 1990 after, by chance, two images among all those taken by a camera during the launch of a rocket revealed the presence of a very ephemeral fluorescence associated with a distant thunderstorm. The first true campaign of observation of these lights, undertaken in 1995 by two physicists[11] from the University of Fairbanks in Alaska, had such large repercussions that, since then, these phenomena have been tracked and recorded a great number of times. This interest was notably used to justify the Lightning and Sprites Observation (LSO) experiment, installed in the international space station in October of 2001. This mission, developed by the French Atomic Energy Commission (CEA) and confided to Claudie HAIGNERÉ, consisted of installing two cameras pointed toward the nadir[12] and triggered during the night when the space station was over a continent where thunderstorms are more frequent than over the oceans. Previous observations targeted instead the horizon and were made notably from terrestrial observatories.

Emitted from the top of clouds, the first lights observed were very bright, blue jets directed toward the tropopause and the stratosphere. They occupy narrow, weakly divergent cones (their half angle at the vertex is about 10°), which come from the troposphere and can attain altitudes on the order of 40 km (fig. 3.9). The speed at which the blue jets grow is estimated to be about 100 km s^{-1}. Much higher, toward the summit of the mesosphere, at altitudes approaching 80 km, red lights can appear very suddenly and disappear a few milliseconds later. They look like shredded strips of cloth that hang toward the stratosphere, sometimes turning like tendrils. These red lights are called *sprites*, and their points can descend to around 40 km of altitude. Finally, above the sprites, at altitudes close to 90 km, other red lights may be discerned that are weakly luminous and that form rapidly expanding circular rings. These red rings are often called *elves*

---

11 This observation campaign was undertaken by Davis D. SENTMAN and Eugene M. WESCOTT from an airplane specially equipped to fly through thunderstorms and photograph the light induced in the upper atmosphere.

12 The nadir is the point in space opposite the zenith on a descending vertical line. It can be observed only from vehicles flying at sufficiently high altitudes.

(for Emission of Light and Very-low-frequency perturbations from Electromagnetic pulse Sources).

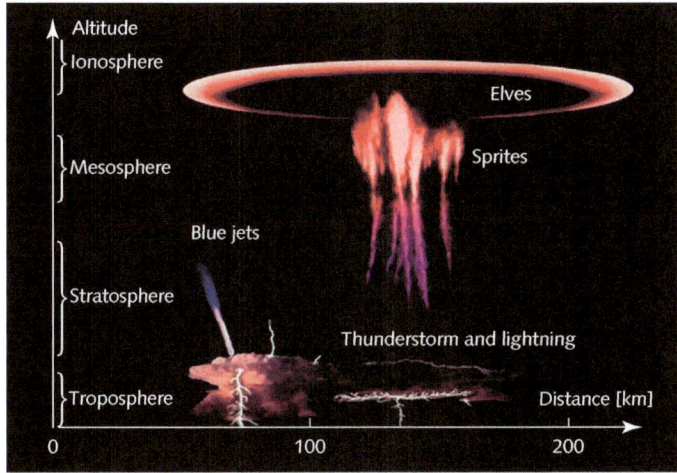

Figure 3.9 - Illustration of the ephemeral luminous phenomena that can appear in the upper atmosphere and that are associated with thunderstorms in the lower atmosphere. From top to bottom, note the elves or halos shaped like expanding red rings, the red sprites leaning toward Earth, a blue jet ascending toward the stratosphere and, finally, the lightning between the cloud and the ground. [© NOAA]

Although these luminous emissions in the upper atmosphere during thunderstorms continue to be the subject of significant research efforts, the acronym elves chosen as their name already indicates that their origin is attributed to electromagnetic impulses generated by an atmospheric storm. As indicated above, the instantaneous magnitude of the electric current in lightning can attain 10 000 A. The magnetic field induced by such a lightning bolt in its immediate vicinity, although not gigantic, is quite significant: on the order of millitesla at 1 m from the lightning bolt. Despite its attenuation as a function of distance, the electromagnetic wave generated by the sudden appearance of the lightning bolt propagates into very distant regions, such as the mesosphere and beyond. The annular form of elves would thus be the intersection between the spherical wavefront that comes from an ensemble of lightning strikes in the lower atmosphere and a plane some 80 to 100 km above the ground, where the ionization of nitrogen makes this luminous emission possible. In fact, nitrogen molecules (N_2), excited by the ionizing perturbations of atmospheric lightning, would be capable of emitting such radiation in the form of wavelengths varying with pressure and, consequently, with altitude. To better know and understand these luminous phenomena, a small satellite called the *Tool for the Analysis of RAdiation from lightNIng and Sprites* (TARANIS) was being developed at the moment that the original French text of this book was being written and began to produce new data in 2013.

2.4. Rainbows

Let us return to the very low atmosphere; within the first tens of meters from the ground. In an air that has become clear but still full of fine droplets, a beautiful rainbow can follow a fearful thunderstorm: calm and gentle after the violence of the debauchery of energy. This colored arc is an optical illusion; it exists only if an observer is present to capture the sunlight refracted towards his retina by some of the fine droplets suspended in the air. Moreover, rainbows are observable not only after a storm but also in all situations wherein the air contains enough fine droplets for the refraction and reflection of the light to generate this effect. Thus, by clear weather, gardeners that water their plants with water-spray systems regularly observe rainbows, provided the droplets are sufficiently fine and numerous. Frequently, tourists can also observe rainbows in the spray above the waves that explode on rocky coasts.

To see how a rainbow forms, we follow a ray of light coming from the Sun. The ray penetrates a drop by traversing the first air-water interface. In so doing, it refracts at an angle that depends on its wavelength. Next, it is partially reflected from the rear surface of the drop and returns towards the front surface where it is again refracted as it crosses the water-air interface. It then continues on to the observer. The theory of these two successive refractions with an intervening reflection shows that the ray that arrives at the observer lies on the lateral surface of a cone whose axis lies on the extension of the line from the Sun to the eye of the observer. The angle between the cone axis and the cone surface varies from 40° to 42°, depending on the wavelength (i.e., the color) under consideration (fig. 3.10).

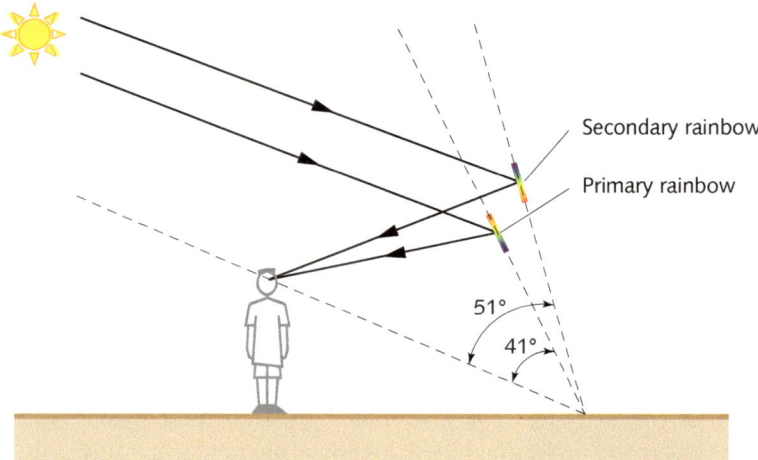

Figure 3.10 - Paths of light rays refracted and reflected by spherical droplets, allowing the observer to observe a rainbow. The primary arc is formed on the cone seen by the observer at an angle of about 40° to 42°, and the secondary arc comes from the droplets on the cone seen at an angle of about 50° to 53°.

Because the angle of refraction varies slightly with wavelength, the various colors do not come from exactly the same cone: when all is done, red lies at the exterior of the arc and blue at the interior. In addition, the cone angle is independent of the size of the droplets, provided that they are spherical, which is the case for very small droplets. In fact, the surface tension of droplets, which is due to capillary action, is the dominant force at work, notably with respect to the weight of the droplet and friction with the ambient air (see insert I3.1). As droplet size diminishes, the shape imposed by surface tension becomes ever more perfectly spherical. If we cannot observe a rainbow at the edge of a dark gray cloud, it is because the drops are too large to be sufficiently spherical and thus the refracted rays are dispersed over a broader range of angles.

Figure 3.11 - Primary rainbow and secondary rainbow, photographed by the author in Grenoble above the Paul MISTRAL Park on April 5, 2012. Note the inversion of the order of the colors between the two rainbows and the increased blurriness of the secondary rainbow.

This circular arc, with red at the exterior and blue at the interior, is the primary rainbow. It is the clearest and often the only rainbow to be seen. But the attentive observer may notice the presence of another arc, called the *secondary rainbow*, which is concentric with the primary rainbow, more blurry, and at the exterior (fig. 3.11). This arc is due to an additional reflection: the ray of light undergoes two partial reflections from the back surface of the droplet before returning toward the front surface, where it refracts as it exits the drop and enters the air. Finally, this doubly reflected ray arrives at the eye of the observer, after two reflections and two refractions. The ray seen by the observer thus lies on the lateral surface of a larger cone seen by the observer at an angle between 50° and 53°.

I3.1 - Surface tension and capillarity

Various experiments show that all interfaces between two fluids, one liquid and one gas or two different liquids, are subject to a tension, such as the membrane of an inflated balloon. As an example, the tensiometer of LECOMTE DU NOÜY demonstrates this phenomenon by measuring the apparent weight of a ring as it is withdrawn from a liquid. When the ring is totally submerged in the liquid, its apparent weight equals the difference between its weight and the buoyant force of ARCHIMEDES. When the ring is completely out of the liquid, the apparent weight equals the weight of the ring. While the ring traverses the free surface, slightly lifting some liquid in the process (see the figure below), the force $\mathbf{F}$ required to maintain it in static equilibrium is greater than its weight by a quantity equal to the sum of the tensile force applied by the free surface all around the contact line with the ring $[2\pi\sigma(R_1 + R_2)]$ and the weight of the liquid lifted. When the radii R_1 and R_2 are nearly equal, the weight of the liquid is negligible and the surface tension can be calculated from $\sigma = F/4\pi R$.

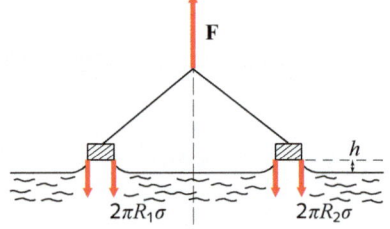

This measurable force implies that the pressure is different on each side of all curved interfaces, such as at the skin of a drop. If p_i et p_e respectively denote the interior pressure (concave side) and the exterior pressure (convex side), the difference between the two is $p_i - p_e = \sigma(1/R_1 + 1/R_2)$, where σ is the surface tension, and R_1 and R_2 are the two principle radii of curvature. For a spherical surface of radius R, this pressure difference reduces to $p_i - p_e = 2\sigma/R$. For a water-air interface at about 15 °C, the surface tension is in the neighborhood of 80 000 N m^{-1}. This magnitude varies slightly with temperature and with the salinity of the water. This pressure applied by the fluid on one side of the interface on the fluid on the other side explains the formation of menisci and, notably, the capillary ascension of liquids such as water or alcohol in small-diameter tubes or in porous media such as soil or absorbent pads. For mercury, which does not wet glass, capillarity causes a depression and the meniscus, which is easily visible in older thermometers, is inverted.

From the relations above, we can easily deduce that the air pressure inside a soap bubble, separated from the outside air by a two surfaces, equals $4\sigma/R$. We also see that, as the radius R becomes very small, the surface tension grows to become the dominant force, overcoming the weight, which diminishes as R^3, and air drag, which diminishes as R^2. Consequently, small drops are spherical in shape, whereas larger ones are deformed by their weight and by air drag. The experiment of LECOMTE DU NOÜY also shows that the additional interface created upon lifting the ring stores the energy that is put into the system. The work done to displace the ring by dx is expressed as $F dx = 2\sigma l\, dx$, where l is the length of the line of contact. This leads to a useful interpretation of surface tension: it is the energy stored per unit area in a gas-liquid interface.

With respect to the primary rainbow, the order of the colors is inverted: red is at the interior and blue is at the exterior. At each partial reflection, a fraction of the incident energy traverses the interface and the remaining is reflected, which is why the secondary rainbow is much less clear and intense than the primary rainbow and often escapes the notice of spectators, being seen only by knowledgeable observers. Searching for a tertiary arc would be illusory.

3. The various precipitations

At various places in the preceding sections, rain and hail have been mentioned, although without giving details of how they form or fall. The first chapter reviewed how the condensation of water vapor is due to the fact that moist air has a saturation vapor density above which water cannot remain in the vapor phase and begins to condense. Once this condition is satisfied, the water molecules dispersed among the molecules of nitrogen, oxygen, and the other components of air can aggregate to form extremely fine droplets.

This stage in itself is not as simple as one might think: the vapor molecules being relatively far from each other within the gaseous mixture, they cannot instantly regroup to aggregate into microdroplets. Forming a microdroplet with a diameter of about 1 μm requires billions of water molecules, whose size is on the order of several angstroms. We can thus understand that their convergence requires time and imposes what specialists call a *supersaturation* or *delayed thermal equilibrium* (see insert I3.2). Apart from this unavoidable delay, another requirement is the presence of microscopic particles on which molecules can aggregate to form the first droplets. These particles are either microdroplets that have already formed or fine dust generally called *aerosols*.[13] These extremely small objects are almost always present in sufficiently large numbers in the atmosphere. They provide the sites to which water molecules attach themselves to aggregate and form first droplets, then drops.

In an analogous fashion, to partially freeze a drop of water and create the first solid crystals, a regular delay is required within each crystallization nucleus (i.e., seed) for the disordered molecules in the liquid state to organize into crystalline networks.[14] Again, starting this stage requires crystallization nuclei; a role that is played by the dust particles held within the liquid drop.

---

13 An aerosol is a suspension in the air of very fine particles that can be solid, such as those from fires, or liquid, such as very fine droplets.

14 The crystalline structure of ice, whose form is fixed by its assembly from prestructured pieces in liquid water, is quite particular and fairly similar to that of diamond. Its cohesion is due to the intermolecular attractions called *hydrogen bonds*.

13.2 - Elements of thermodynamics applied to air and water

A body is in equilibrium when it exchanges nothing with its environment: neither mass, momentum, nor energy. The set of quantities that characterizes its state; in other words, pressure, temperature, and all its physical properties such as viscosity, heat capacity, and conductivity, are called *state variables*. In the case of a homogenous body, these variables are determined by only two among them, so that the rest may be calculated from these two principle variables via relations called *equations of state*. Normally, temperature and pressure are chosen as the principle variables. Thus, the density of a fluid and its viscosity are often expressed as functions of pressure and temperature. For a mixed body, with each addition of a new species, the concentration of this species suffices to characterize the new mixture. In a binary mixture, the three principle variables are the pressure, the temperature, and the concentration of one of the species.

Sufficiently dilute gases, such as air, follow fairly well a very important approximation that is called *the ideal gas law*. This law describes ideal gases, for which the equation of state is $p/\rho = RT/M$, where p is the pressure, ρ is the density, $R = 8.314472$ J mol^{-1} K^{-1} is the universal gas constant, T is the temperature, and M is the molar mass of the gas. The universal gas constant R is related to BOLTZMAN's constant ($k_B = 1.380649 \times 10^{-23}$ m^2 kg s^2 K^{-1}) and to AVOGADRO's number ($N_A = 6.0221415 \times 10^{23}$) by $R = N_A k_B$. For air, despite it being a mixture of various species, this approximation remains adequate provided M designates the mean molar mass of the mixture ($M = 0.028964$ kg mol^{-1} for dry air).

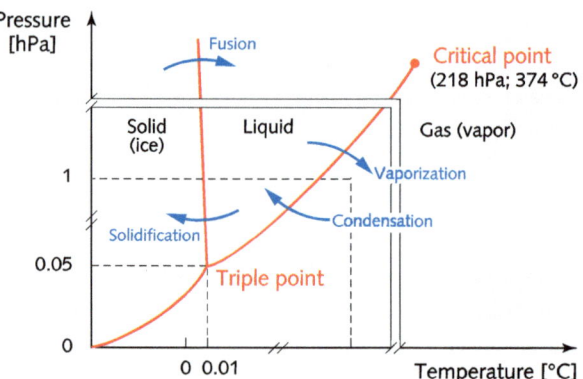

Phase diagram for water. The curves indicated by arrows represent the typical trajectory taken by a water particle upon melting, evaporation, or freezing. Note that the critical point of water is difficult to attain (the horizontal and vertical axes of the graph are interrupted).

Energy must be supplied to or removed from a mass of matter to change its state. This change satisfies the first law of thermodynamics, which says that the work done by external forces applied to a system plus the thermal energy supplied to the system equals the change in internal energy of the system. In practice, this internal energy is just the sum of the energy of each elementary particle; molecules in the case of a gas. Although matter is rarely in equilibrium, its changes of state and its motion are generally extremely slow compared to the motion of molecules or atoms. This justifies the local-equilibrium approximation, which assumes that, at all times, all portions of a system that move slowly with respect to the mean speed of its constituent particles are very close to equilibrium, and that any phase transitions within the system can be described as a sequence of equilibrium states. In this case, we can generalize the expression of the first law as follows: the sum of the work and thermal energy supplied to the system equals the sum of the change in its internal energy and the change in its kinetic energy of macroscopic motion. This generalized form of the first law is what we apply to air, to water, and to all their interactions; notably, the vaporization of water in air, or the condensation of water into droplets or on cold surfaces.

Phase transitions can be represented on a diagram that, for a homogenous system, appears as a two-dimensional graph, analogous to that shown in the preceding figure, which considers the case of water. All points on this diagram represent an equilibrium state for a mass of water. All changes in state, which are taken to be sequences of equilibrium states, can thus be represented by a trajectory on the diagram. For water, each phase (solid, liquid, or gas) corresponds to one of the three regions separated by the curves that intersect at the triple point: the solid state at low temperature and the liquid or gas state at higher temperatures. The trajectories represented by the blue arrows correspond to the phenomena of vaporization or condensation or of melting or freezing. On the figure, the critical point (218 hP, 374 °C) is indicated just for completeness; it is not attainable in the atmosphere or in the oceans.

In each particular case, the work done is expressed in terms of the relevant forces and the displacements of the points at which these forces are applied. Likewise, the thermal energy supplied is expressed by using the law appropriate for the given form of heating (the STEFAN-BOLTZMANN law for radiation, the FOURIER law for conduction). For water, we should also include the contributions from the latent heat at constant temperature of vaporization (2257 kJ kg^{-1} at 100 °C, and the opposite for condensation) and of melting (333.55 kJ kg^{-1} at 0 °C, and the opposite for freezing).

The interpretation in terms of energy also leads to a molecular interpretation of surface tension. Imagine a single molecule near an interface between two fluids. In creating this interface, we removed all molecules of the same species within a half space on one side of the given molecule and replaced them by molecules of the other fluid. The surface tension is the net energy supplied for this double operation.

When the atmospheric pressure drops below the saturation vapor pressure, the first microdroplets formed are too small and light to precipitate. On their scale, the forces linked to volume, such as weight, are proportional to the cube of their diameter and are negligible in comparison to the forces related to surface area, such as the drag in the ambient air, which are proportional to their diameter squared. Thus, the microdroplets remain suspended in the air. They can whiten the air or make it opaque if they are sufficiently numerous, even if the wind is extremely weak or almost nonexistent. This is how fog forms, which is easily seen in winter above cold ground or water and which can last for hours despite weight acting on each individual droplet.

Depending on the weather conditions and on the speed with which pressure and temperature change along their trajectories, these suspended droplets can grow by aggregating with other molecules and by coalescing with other droplets or, on the contrary, they can release water molecules back into the surrounding gaseous phase. The relative contribution of evaporation with respect to condensation explains how clouds manage to disappear without precipitation. Precipitation in the form of rain occurs when the balance of forces tips in favor of the weight, which implies that the drops have attained sufficient size. We can also add that, even before microdroplets form in a large volume of air, condensation often appears at the boundaries, on the coldest surfaces. This surface condensation leads to the dew in the prairies or to the fog on the inner surface of a window pane with a colder exterior surface.

With respect to thunderstorms, we discussed above convection cells such as in cumulus or cumulonimbus clouds, within which the central updraft constantly lifts droplets, even those that would precipitate in calm air. We noted that precipitation occurs on the periphery of the cloud, in the downdrafts that they drag with them. Some drops may circulate several times around such a cell before they precipitate, meeting and aggregating with many other drops and growing quite large, making them precipitate in the end. All along this cloud-bound trajectory, drops endure alternating episodes of melting at the bottom where the temperature is often above 0 °C, and freezing at the top where the temperature is much less than 0 °C.[15] These changes of state can involve the entire drop if it is small enough. However, when the radius of the drop becomes sufficiently large, the boundary between melting or freezing does not have time to cross the entire drop. In this case a structure of concentric shells appears, formed due to the alternating episodes of freezing and partial melting endured by the drop during the climbs and plunges. This onion-peel structure is rather characteristic of large hailstones; those that surpass several centimeters in size. Their impact on vegetation, garden greenhouses, automobiles, and rooftops are capable of causing serious damage.

The formation of sleet in winter is completely different. Raindrops, which are liquid when they begin their descent, freeze on their surface if they pass through a layer of cold air before arriving at ground level, where the temperature is less than 0 °C. The liquid water within the thin ice crust is often in a supercooled state (another example of delayed thermodynamic equilibrium). Its temperature is already less than 0 °C, but the water molecules have not yet had time to settle into a solid

---

15 In the atmosphere, the temperature drops 6.5 °C per kilometer but, in stormy conditions, the intense turbulent kneading can reduce this drop to around 5 °C per kilometer. No need to look further to see that the temperature difference between the top and bottom of a cumulus cloud 5 km high is on the order of 25 °C, which explains the quasi-systematic freezing of drops that pass through the top part of such a cloud.

crystalline network. In general, the small pellets of sleet burst upon hitting the ground, and the supercooled liquid water that escapes freezes very rapidly, often forming sheets of black ice.

The formation of black ice after sleet burst open on the ground is not the only cause behind this phenomenon. Surface ice also forms because vapor directly condensates into the solid state on frozen ground. The formation of surface ice often passes through an intermediary step in which a layer of frost, which is a sort of partially frozen dew, is deposited on plants or other surfaces. At the end of winter, the formation of frost over buds protects them from internal freezing, which would cause the plant fibers to burst and thereby compromise any subsequent flowering. When rain falls on frozen ground, the liquid water, which is often in a supercooled state because it has not had time to solidify, contributes by thickening the surface ice.

We may now wonder how snow forms. At the outset, inside a rather cold cloud, condensation appears spontaneously on the aerosols in the atmosphere, changing the state of matter from gaseous to solid without passing through the liquid phase. The crystals are initially miniscule and aggregate to form beautiful dendritic structures, or stars. As they begin to form, these firstfruits of snowflakes are so light and, because of their fine geometry, their surface area is so large that they float in the air, barely visible and unable to precipitate. In the mountains, these embryonic snowflakes may be seen scintillating in the sunlight above a layer of fresh powder snow. Given the proper conditions, each of them can serve as seeds around which others can aggregate, making the seed grow into bona fide snowflakes that can precipitate. In the lowlands, a wind heavily charged with these new flakes can make a blizzard, as is fairly common in the Midwestern American states, given the cold air coming from the north.

The snow starts to fall once the balance of forces is tipped in favor of weight. Frequently, during their descent, snowflakes grow larger by grouping together. Depending on the local conditions, they can conserve their state and their initial finely branched morphology, which creates powder snow; a frequent occurrence at high altitude. If the temperature is slightly higher, they may melt partially on their surface. This superficial melting tends to smooth their initial form, and they grow larger as they aggregate with other flakes and with liquid drops. This creates heavy snow, more frequent at lower altitudes under conditions conducive to an intermediate state where snow and rain may become mixed together.

4. How are weather forecasts produced?

The difficulties of weather forecasts were briefly touched upon several times in this chapter and the preceding. The stakes involved are considerable and condition entire swathes of the economy, notably in areas such as agriculture, fishing, tourism, air transport, and public health. We shall try to more precisely define this challenge.

The evolution of weather conditions is currently calculated by using high-precision numerical simulation techniques that involve refined mathematical models implemented on powerful computers that are much faster than nature itself: in several hours, these numerical simulations can predict the weather for the following days based on initial conditions supplied by dedicated satellites. These simulation techniques proceed by slicing up the troposphere over a large horizontal distance (thousands of kilometers, or even tens of thousands) and into a fine, virtual mesh. The evolution of each cell of the mesh is governed by the equations of fluid mechanics, which express the balance of all the relevant magnitudes: mass of each constituent, water and carbon dioxide included, momentum, and energy. The equations are completed by the boundary conditions, which model the interactions between the piece of the troposphere under consideration and the ground and oceans underneath, and the tropopause and outer space above, without forgetting the input due to radiation, which is positive during the day due to sunlight and negative at night due to the radiation emitted by Earth toward space. The initial conditions supplied by the satellites are vital data for initiating the calculation. The resolution of the gigantic system of equations begins at the initial time and, by calculating small, successive temporal increments, it targets successive instants in the future. The end product is what can be called the *meteorological trajectory* of the domain considered.

With their spatial limitations, the equations and conditions necessarily involve approximations, but the errors introduced are known and predictable or can be reduced if results of greater precision are desired. In contrast, the initial conditions present a considerable challenge. These are, in fact, well known on all scales for which satellites can detect the properties of the atmosphere. But they are not known on scales smaller than the resolution of the satellite observations, such as for shutters that bang and or for birds that fly off. However, this uncertainty on the small scales can progressively transfer to larger and larger scales, until it attains a planetary scale after about ten days. This is a fundamental property of the equations of fluid mechanics that has been well established since the 1980s and is often illustrated by what is called the *butterfly effect*. No matter the effort dedicated to improving the resolution of the satellite measurements, scales exist that are smaller than the smallest scale resolved with precision, such as the scale of butterfly wings. Thus, the uncertainty initially concentrated in the small scales inevitably ends up contaminating the large scales, both in reality and in calculations.

Of no significance on short timescales, this uncertainty becomes an enormous challenge after several days. To evaluate it, the following exercise was done: starting from two identical sets of initial atmospheric conditions, one was modified by adding a random perturbation of very weak amplitude and on the very small length scales. The results of these numerical simulations showed that, as more time goes by, the difference between the two simulations becomes greater and greater, with the two predictions becoming less and less similar. After about ten days, one of the simulations predicted a high-pressure zone in a region where the other predicted a low-pressure zone. Thus, this exercise shows that, because of the butterfly effect, deterministically predicting the weather more than ten days in advance is not possible.

This time limit for deterministic predictions is linked to the typical values of the length scales and the speeds of depressions, which is linked in turn to the value of the REYNOLDS number for the troposphere. This challenge is not restricted to meteorological predictions; it is universal and affects all numerical predictions involving turbulent flows. Nonetheless, when the goal consists of predicting a flow over the short term, such as the transit of air from one end to the other of the fuselage of an airplane in flight, for example, the effect on the results is of no consequence because the uncertainty does not have sufficient time to contaminate the relevant scales. To keep things simple, let us say that the time beyond which such predictions lose their validity is of the order of magnitude of the characteristic time of the largest scales of the flow. In practice this pertinent timescale is the time it takes for several revolutions of the largest vortices, or the time required for the largest depressions to make several complete revolutions. In the preceding chapter, we found that an average depression took some fifty hours, or about two days, to complete a single revolution. This is consistent with the timescale of ten days required for two sets of initial conditions, supposedly identical, to evolve into two entirely different scenarios.

Do alternative methods exist? Of course, meteorologists also make statistical studies consisting of identifying typical sequences, with some interesting results. But, whatever the procedure, the conclusion is a probability—no more, no less. Even if the study involves numerous cross-checks, even if it is very refined, and even if the specialists know how to extract very useful lessons, this statistical prediction remains of the same stripe as observations based on the gardeners frog or on well-known proverbs.[16] Everyone knows that probabilities are an excellent tool within the framework of the law of large numbers: if we repeat a sequence a great many times, on average the most probable event is the one that occurs. But the farmer who wants to know if a freeze will occur within the next two weeks is not at all within this realm, because she is interested only in a single sequence; namely, the next two weeks. Her case would systematically fall outside the most probable evolution simply because it is unique. In conclusion, statistical methods are of interest for those concerned with the average tendencies, but they do not help those who need to know a specific meteorological sequence.

An example would perhaps better illustrate this conclusion. If the manager of a ski area wants to know the average snow depth of the area over a significant number of winters, the statistical method will provide completely reliable data. In contrast, if in November he wants to know the snow depth for the coming winter, the dates

---

16 Numerous examples exist of these proverbs that contain a grain of truth, but on which it would be unreasonable to base a prediction of a specific event: *Should Saint MÉDARD'S day be wet, it will rain for forty yet; at least until Saint BARNABAS, the summer won't favor us.*

of the strongest precipitation and those of the warmer periods that would partially melt the snowpack, he would be misadvised to put faith in the statistical average or in any other method of prediction. This represents one of the profound limitations of weather forecasts.

Conclusion

In all the phenomena described in this chapter, water in the air is of primal importance, much more than the other constituents because, under atmospheric conditions, it can morph into its various states: gaseous, liquid, or solid. Its condensation into droplets forms the clouds and serves as a sort of maker for depressions. Its presence is revealed by phenomena of an extreme variety; either very calm such as dew and fog, or very violent such as thunderstorms and tornados, after which may follow a rainbow announcing the return of clear times.

The physics behind the phenomena described and observed by everyone in the lower atmosphere is relatively clear, even if some phenomena, such as ionization in clouds during thunderstorms, still pose serious challenges. The equations of fluid mechanics on which are founded meteorological models can be resolved with good precision. Nonetheless, deterministically predicting the weather two weeks hence remains impossible. This is not because the equations are ill conceived or solved with too little precision. Rather, it stems from the fact that turbulent flow, notably in the troposphere, is extremely sensitive to the initial conditions: a very small error in the initial conditions, which in reality comes from an inevitable uncertainty on the very small length scales, can generate after a finite time major errors across all scales.

The most intermittent phenomena, such as thunderstorms, extend in spectacular ways into the high or very high atmosphere, which is much less well understood. The observation of these phenomena is still recent, and their ephemeral traits have led them to be named after whimsical fairies, the sprites, or mysterious geniuses, the elves.

Chapter 4

Heavier than air, how can they fly?

The wings quivered under the evening breeze
With its song the engine rocked the sleeping soul
The Sun brushed us with its pale color.

(Antoine DE SAINT EXUPÉRY, *Wind, Sand, and Starts*)

© Springer International Publishing AG 2017
R. Moreau, *Air and Water*,
DOI 10.1007/978-3-319-65215-3_4

A dvancing our knowledge of the atmosphere was not done by just observing it from the ground: it had to be crisscrossed and probed with the appropriate vehicles, such as the planes or weather balloons used by the meteorological authorities. The observation of birds and insects proved that it was indeed possible to make ponderous objects fly, without having them fall like rocks. That said, since the dream of Icarus[1] and the first attempts of Leonardo da Vinci,[2] what efforts were not required to first understand how an object heavier than air could fly and then to construct the marvels of technology that are modern airplanes? Let us first focus on analyzing the origin of this force called *lift*, which seems capable of vanquishing weight. Next, because we shall discuss airplanes, helicopters, and wind turbines, we will also attempt to explain the origin of one of their major drawbacks: the sometimes deafening noise that they emit. Finally, we shall try to understand how and why the diffuse noise from a flying airplane can generate such deafening shockwaves when it flies at supersonic speeds, which will bring us to question the reality of the sound barrier, which is often evoked by pilots.

1. Lift and drag

1.1. Lift explained by pressure forces

The combination of Earth's gravity and of a mass implies that every object in the atmosphere, notably every part of the atmosphere itself, is attracted because of its weight toward the center of Earth. At rest, all objects in the atmosphere are also subject to a lift from the surrounding fluid, which is called *buoyancy*, as first expounded by Archimedes' principle.[3] In the calm air of a high-pressure zone, these two forces balance each other everywhere, and no movement appears. But as soon as an object is heavier or lighter than the ambient air, or if a variation in the local density of a portion of the air itself modifies the balance between these opposing forces, the equilibrium is lost and motion ensues. This is the case, for example, when a chimney ejects warm smoke that is less dense than the surrounding air,

---

1   The first record of the human dream of flying like the birds dates from Ancient Greece. In the mythology of the times, Icarus and his father Daedalus dream of flying to escape their exile on Crete and return to Athens.

2   The painter, philosopher, and Renaissance scholar Leonardo da Vinci (1452–1519) was fascinated by the flight of birds and imagined numerous flying machines. Most were not realistic, such as his aerial screw (i.e., helicopter), but the hang glider can be considered to be a sort of replica of some of this drawings.

3   Archimedes of Syracuse, who lived in Sicily in the 3rd century BC, is considered as one of the greatest scientists of classical Greek antiquity.

thus causing the smoke to rise from the chimney. As soon as the smoke cools by mixing and exchanging thermal energy with the ambient air, its ascension ceases because buoyancy and weight cancel each other. Such is also the case, but in the opposite sense, when an object heavier than the ambient air falls, such as a rain-drop, a hailstone, or an overripe fruit.

However, this is not the case for a bird or airplane in flight, for these are subject to a third force, lift, which can overcome the difference between weight and buoyancy. It is also not the case for a javelin propelled almost horizontally through the air. In this case, the thrower gives the javelin an initial momentum (see insert I2.1), which allows it to advance through the air in spite of friction acting all along its surface. This example highlights another force that we shall discuss; namely, drag, which is the resistance by fluid to the motion of an object within it and which is composed of the friction exerted by the ambient air all along its surface and the force of pressure that acts on the leading edge, itself connected to the effort required to separate the fluid layers to allow the object to penetrate the air. The origin of these forces, which is indispensable for a proper understanding of the dynamic phenom-ena observed in air and in water, merits some discussion.

Let us return briefly to weight. It reflects a very general phenomenon, known since GALILEO,[4] even if others such as NEWTON and EINSTEIN have since refined the law: the universal attraction of bodies. In fact, all bodies attract each other by a force that is proportional to the mass of each body and inversely proportional to the square of the distance between them.[5] On our planet, the attraction of solid bodies by Earth itself thoroughly masks their mutual attraction and gives us the acceleration of gravity, which is 9.81 m s^{-2} at sea level (about 6300 km from the center of Earth). Considering the thinness of the troposphere (8 to 15 km), the variation of gravity as a function of altitude in the low atmosphere is very moderate; at most of the order of 0.15%, although this does not mean that it has no effect, as we shall see in the next chapter on the topography of the free surfaces of the ocean.

Buoyancy itself is a result of weight. All bodies plunged into a fluid at rest, such as air or water, and that are delimited by a closed surface S, are subject to a buoy-ancy force of equal magnitude as the weight of the fluid displaced but that acts in the direction opposite the weight (i.e., upward). We can convince ourselves of this by the following argument: when the surrounding fluid also fills the interior of the closed surface S, it is at equilibrium. The buoyancy of the fluid surrounding the surface S thus opposes exactly the weight of the fluid within the surface S.

4   GALILEO (1564–1642) was an Italian physicist and astronomer and one of the founders of classical mechanics. He contributed significantly to the acceptance of the heliocentric theory, according to which Earth revolves about the Sun, as opposed to the geocentric theory that prevailed in his day.

5   The attractive force between two masses m and m' separated by distance d is expressed as Gmm'/d^2, where $G = 6.674 \times 10^{-11}$ m^3 kg^{-1} s^{-2} is the gravitational constant. From this we can deduce the acceleration due to gravity is $g = Gm_T/d^2$. Given that $g = 9.81$ m s^{-2}, the mass of Earth is found to be $m_E = 1.9 \times 10^{24}$ kg.

Now, without modifying anything exterior to the surface S, we imagine the surface enclosing another object, the buoyancy force remains unchanged. This object is thus subject to a force of equal magnitude as the weight of the fluid displaced but that acts in the opposite direction. The buoyancy force on a solid object moving through air must in general be negligible because air is about 1000 times less dense than such objects. In contrast, it is not negligible at all in water, which is 800 times more dense than air, or almost as dense as many solid objects.

Lift is more difficult to understand because it results from the motion of the object. To explain its origin, consider first an extremely simplified case by imagining a thin, symmetric, infinitely long wing. The flight of this wing at a speed V becomes a two-dimensional phenomenon, as illustrated in figure 4.1. Imagine an observer fixed to this wing who, like an airline passenger, sees the external fluid come from upstream at a speed V. In case **(a)**, where the angle of attack of the wing is zero, the flow above the wing, or along the extrados, is identical to the flow underneath the wing, or along the intrados. The pressure on the intrados and extrados is fairly symmetric so the vertical components can only cancel: this wing experiences zero lift. In contrast, in case **(b)**, where the angle of attack of the wing is nonzero, the symmetry is broken, as indicated in figure 4.1. The airspeed is significantly greater along the extrados than along the intrados.[6] To slow the fluid particles that pass along the intrados, they must encounter some resistance. The resistance that slows their motion is due to the overpressure that occurs along the intrados, which locally slows the air by pushing it downward. The result is that the pressure on the intrados exceeds that on the opposite face, so that the net pressure over the entire wing surface gives rise to a net force with a nonzero vertical component: the wing thus experiences a lift, which we can see is linked to the angle of attack.

Figure 4.1 - Cross section of an infinitely long symmetric wing in a uniform wind coming from the left **(a)** with zero angle of attack, **(b)** with a nonzero angle of attack. Note that the flowlines (lines tangent to the velocity) are significantly farther apart on the intrados than on the extrados.

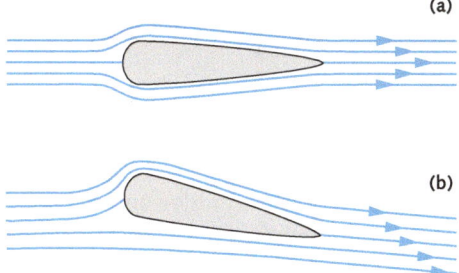

(a)

(b)

For an airplane wing, the angle of attack is set by the pilot who, to take off, extends the flaps while the airplane rolls down the runway. Even if she does not alter the angle of attack of the entire wing, the pilot breaks the symmetry between the intrados and the extrados, which generates sufficient lift to overcome the weight.

---

6   At a fixed altitude, the variations in pressure p and speed V along a given flowline (everywhere tangent to the velocity) are related by Bernoulli's law $p + \rho V^2/2 = const$. The result is that the pressure is maximal where the speed is minimal, which is to say at the stagnation point, and vice versa.

Inversely, to descend before landing, the pilot reduces the angle of attack by raising the flaps, thereby reducing the lift or even making it negative so that it adds to the weight. For birds, the animal itself exerts its muscles to create lift.

The fact that a wing has a finite thickness reduces the efficiency of this mechanism, but does not compromise the principle. In aeronautics, the word *thinness* is used to signify the ratio of wing thickness to length. Boat sails constitute good examples of thin wings; once inflated by the wind, they curve, which accentuates the dissymmetry and increases the net pressure, leading to sort of horizontal lift that is transmitted to the boat by the mast. The combination of lift from the sail and that exerted by the water on the rudder (another thin sail) and eventually on the keel (a third thin sail) allows one to even sail upwind. The rear spoiler of race cars is also conceived to generate lift, although negative sense; in other words, directed down toward the ground like weight to improve handling.

1.2. Formation of vortices around a wing in flight

To better explain lift as the sum of pressures, a common approach is to substitute a very interesting and strictly equivalent variation that consists of representing the airfoil by a vortex. To introduce this notion and show that the thickness can be neglected, consider first a circular cylinder of infinite length placed in an airstream coming from the left (fig. 4.2).

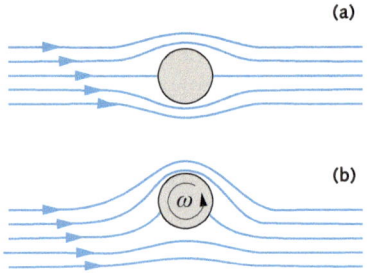

Figure 4.2 - Cylinder in wind coming from the left: **(a)** nonrotating cylinder, **(b)** cylinder rotating with angular velocity ω.

In figure 4.2a, the intrados and extrados are perfectly symmetric, so the net pressure is zero and the lift is zero. In contrast, when the cylinder rotates about its axis as in figure 4.2b, the two stagnation points shift downward and the streamlines passing under the cylinder become less dense, making the local speed diminish and the local pressure increase. Conversely, the streamlines passing above the cylinder become denser, so the local speed increases and the local pressure decreases. The net force due to pressure over the entire surface of the cylinder leads to a nonzero lift that is proportional both to the intensity of the vortex modeled by the rotating cylinder and to the speed of the far-upstream airflow. In fact, rotating masts analogous to the rotating cylinder of figure 4.2 have been designed to power ships, although their implementation failed due to the fact that, in the case of high winds, they are much harder to retract than conventional sails.

Imagine now that, instead of the whole cylinder rotating, only a thin film rotates on the surface—a sort of skin wrapped around the contour of the cylinder. For the flow of the ambient fluid, which feels only this skin, the result is the same. The cylinder with this moving skin is subject to a lift whose sign depends on the direction of rotation of the skin. Moreover, it can be shown that the lift is proportional to the density of the fluid, its speed far upstream, the speed of the skin, and the perimeter of the circle[7] or of the flexible contour.

The analysis is not finished, because the wings of airplanes are not covered by moving skins. In fact, what matters is not the speed of the points along the surface of the wing or cylinder, but rather the speed of the fluid just external to this very thin fluid layer, which is called the *boundary layer*, (already introduced in chapter 2, insert I2.3). The boundary layer around a wing is outlined in figure 4.3, where its thickness is greatly exaggerated to make it visible. Note that it begins at the leading edge, where it is extremely thin, and becomes progressively thicker on each side of the wing.

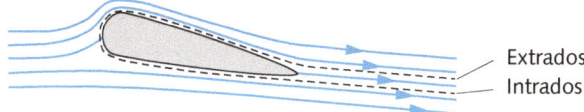

Extrados
Intrados

Figure 4.3 - Wing from figure 4.1b, showing the intrados and extrados boundary layers that meet at the wake of the wing after having been subjected to different variations in thickness along the intrados and the extrados.

Because of the angle of attack, it is not symmetric, and the intrados and extrados layers both recombine at the trailing edge of the wing to form the wake of the wing. For example, if the distance from the leading edge to the trailing edge is of the order of one meter and the speed is of the order of 100 m s^{-1} (or 360 km h^{-1}), the thickness of this boundary layer is less than 1 mm at the trailing edge, where it is at a maximum. If the angle of attack of the wing is positive, the boundary layer becomes significantly thicker below the wing, where the REYNOLDS *number* is relatively small, than above the wing. Because of this asymmetry, the velocity averaged around the closed contour that traces the exterior of the boundary layer is nonzero, as if the wing and the boundary layer formed a moving skin that was made to rotate from the interior. This averaged velocity is proportional to the

7   This is given by the KUTTA-JOUKOWSKY theorem, which is expressed as $P = \rho V_{fl}\Gamma$, where P is the lift per unit length of the cylinder, V_{fl} is the fluid velocity far upstream with respect to the cylinder, and $\Gamma = 2\pi r^2 \omega$ is the circulation of the fluid about the cylinder (the product of the tangential speed ωr and the circumference $2\pi r$). The circulation about a closed contour encircling an object is given by the flux of the vector $\nabla \times V$ across the surface delimited by this contour and is an integral quantity proportional to the velocity averaged around the contour. Its analysis is far beyond the scope of this book. The vector $\nabla \times V$ is often called the *vorticity*. This theorem, which can be generalized to account for various two-dimensional forms of wings, is attributed jointly to the Russian physicist Nikolaï JOUKOWSKY and to the German mathematician Martin Wilhelm KUTTA, both of whom contributed significantly to the development of aeronautics in the early 20th century.

circulation of the velocity vector around the closed contour of this ensemble, which is equivalent to the intensity of the vortex, itself created artificially in the case of a rotating cylinder.

To recap, a vortex appears during takeoff as soon as the pilot gives the wings a nonzero angle of attack, which provides lift. In addition, an opposite-rotating vortex is left at the departure point because, initially, no vortex exists and this quantity is constant on timescales that are short with respect to the characteristic time of viscous friction. Thus, by creating a vortex that gives rise to lift, the pilot also creates an opposite-rotating vortex that remains at the airport. Based on the equation of fluid mechanics, the two explanations of lift (i.e., net pressure and generation of a vortex about the wing) can be shown to be strictly equivalent.

One step remains to be taken to arrive at a proper understanding of lift; namely, to go from an infinitely long wing to one of finite length. Consider figure 4.4, which illustrates qualitatively, without respecting the length scales, how a vortex evolves along a wing and cancels at its extremities. The slowing of the fluid particles and the overpressure on the intrados both are at a maximum at the midpoint of the wing and cancel at the extremities.

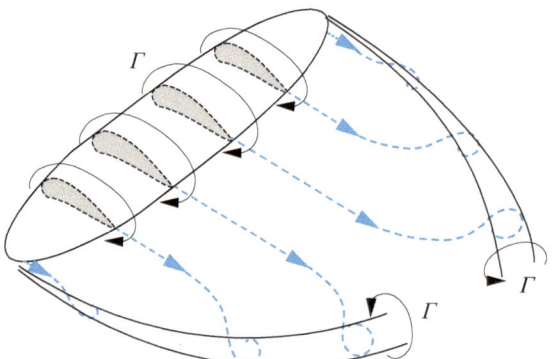

Figure 4.4 - Schematic illustration of the airflow around a finite-length wing with a circulation Γ that is constant along the large vortex tube that forms the wing-tip vortices—themselves closed by the vortex tube left at the airport (not shown). Note the direction of the vertical airflow induced by the wing-tip vortices: ascending at the exterior and descending at the interior.

A perfectly analogous situation occurs on the extrados, but of opposite sign. This effect thus varies along the length of the wing: maximum at the midpoint of the wing and weaker upon approaching the extremities, where zero effect is imposed by the requirement that the pressure be the same whether we arrive along the extrados or along the intrados. The large overpressure on the intrados pushes air from the midpoint toward the extremities, whereas the depression on the extrados attracts air from the extremities toward the midpoint. We thus obtain a component of air velocity along the wing length, going from the midpoint of the wing toward the extremity on the intrados and in the opposite direction on the extrados. These transverse flows lead to the creation of a vortex layer at each extremity of the wing, thus forming two so-called *wing-tip vortices*, as shown in figure 4.5. This structure closes upon itself by virtue of the vortex that remains behind at the departure point and of the vortex created around the wing, thus forming a long vortex tube closed upon itself and along which the circulation of the velocity, or vortex intensity, is constant. In figure 4.5, the condensation of water vapor within the depression inside these vortices created on the flap tips renders the vortices visible.

Figure 4.5 - Demonstration of wing-tip vortices due to the condensation of water vapor at the center of the vortices, where the pressure is below the ambient pressure so as to balance the centrifugal force. In this particular case, where the airplane is preparing to land, visible contrails appear both on the extrados (above the wing) and at the extremities of the flaps, which provide negative lift. At the bottom of the image, note the large spiral vortices, which are due to shearing of the layers of air above and below: this shows that the atmospheric pressure is close to the saturation vapor pressure. [Boeing 757-2B7 © Steve MORRIS]

Do not confuse the long cloudy tube that may be present on the extrados of the wing with those running along the axis of the wing-tip vortices, which, along with the long contrails, are easily visible in the wake of jet airplanes. The former is visible, as in figure 4.5, only if the nearby atmospheric pressure is sufficiently close to the saturation vapor pressure for the centripetal depression of the wing-tip vortices to render visible the condensation of the water vapor dissolved in the air. The latter, on the contrary, are due to the condensation of significant amounts of water vapor coming from the combustion and ejected by the rear nozzle of the jet engines and are the classic signature of the passage of a jet airplane; they are always visible, even in a robust high-pressure zone.

The V formation of flocks of migratory birds stems from the fact that each of them, except the first, places itself in the ascending airflow of the wingtip vortex of the preceding bird, thereby reducing the effort required to maintain altitude. All objects that are subject to lift, be they airplanes, birds, or sails, must therefore be associated with the formation of a large vortitial structure in the form of a long tube that closes upon itself. This tube lengthens during flight and its vortex intensity is linked to the lift. It kinetic energy grows incessantly and is withdrawn from the kinetic-energy budget of the objects in flight.

Balls and balloons are examples of very small flying objects. The lift due to a vortex can explain their sometimes surprising trajectories, such as those of a tennis ball or a ping-pong ball, which are more or less curved depending on the spin applied by the racquet. When the racquet passes over the ball, the trajectory is strongly concave downward. The ball is said to have *topspin*. When the ball is sliced, it has underspin. The curved trajectory in the horizontal plane of a soccer ball, which can allow a corner kick to enter directly into the goal, is another example. Note that the racquet and the fuzz greatly facilitate the generation of rotation for tennis balls, whereas generating significant curve by kicking a soccer ball requires a rarer talent.

The role of viscosity was briefly evoked with regard to the thickness of the boundary layer; we must return to this subject now to answer another question: how long lasts the vortex left behind at the airport by an airplane? On the scale of this large structure, the viscosity is negligible. Yet, it is the only mechanism that can dissipate the kinetic energy of the vortex. Its action, which is quite relentless, works at its own pace, after the initial vortex undergoes a whole series of transformations that pulverizes it into smaller and smaller vortices, each of which in turn pulverize themselves into even smaller vortices, in a cascade process that carries the initial kinetic energy to quite small scales where the viscosity can act on them and transform their energy into heat. This energy cascade, which constitutes one of the profound characteristics of turbulence, is discussed in the appendix. For now, note only that it requires a long time, precisely because the viscosity of air is very small; to the point that an airplane that leaves from Paris can land in London before the kinetic energy of the vortex left in the Paris airport completely dissipates into heat. These vortices left at all airports by all airplanes that depart, one after the other at a high frequency, explains the strong and turbulent winds that sweep through these large spaces and the neighboring fields.

Now that we have introduced the viscosity of ambient fluid and its first manifestation in the form of a thin boundary layer around the wing, introducing friction follows somewhat naturally. Locally, each square centimeter of the wing surface is subject to both the force of pressure (already evoked), which acts perpendicular to the surface, and a tangential force due to friction. The latter is proportional to the fluid viscosity and to the gradient[8] of the velocity across the boundary layer. It acts along the local tangent to the wing contour and in the direction opposite that of

---

8   The concept of gradient, which we already evoked, is very general and is vital for analyzing phenomena characterized by nonuniform distributions of physical quantities, such as the fluid velocity across the boundary layer, or the pressure and temperature in the troposphere. To simplify things, take as a given that the gradient of the fluid velocity across a boundary layer of thickness δ is the ratio V/δ. To be more precise, this gradient is a vector oriented toward the highest speeds, and its magnitude is the derivative of the speed with respect to perpendicular distance from the wing.

the velocity. Overall, on the scale of the entire wing, these friction forces add up and contribute to slowing the movement of the wing through the air.

However, the drag, or the resistance to motion of the object, cannot be reduced to only this first part, often called *friction drag*. It also includes the horizontal component of the net force due to pressure, which complements the vertical force that we have called *lift*. In fact, near the leading edge of a wing, or of any flying object, an overpressure appears that separates the fluid layers whose place is being taken by the object. This overpressure is a maximum at and near the leading stagnation point and decreases as we approach the trailing edge of the wing. There, around the trailing stagnation point where the wake forms as the intrados and extrados boundary layers reunite, it is not balanced by an analogous overpressure. This explanation suffices to make a first approximation of pressure drag: it is the product of the pressure at the leading stagnation point ($\rho V^2/2$), which can be easily deduced from BERNOULLI's equation, and the midship section[9] of the wing. The importance of the wing's thinness (inversely proportional to the midship section), or the aerodynamic profile, then becomes clear. We can also note that the two parts of drag come from different laws, one that is independent of viscosity and proportional to the air density, the square of the air velocity, and the midship section, and the other that is proportional to the viscosity, the air velocity, and the area subjected to friction.

It remains to be understood how an airplane or bird can overcome drag. For this, we must call upon NEWTON's second law of motion,[10] according to which the sum of the forces applied to a body in motion equals the time derivative of its momentum (where momentum is its mass times its velocity). Drag, as small as it might be, is always antagonistic and slows the motion, which can only be sustained if another force exactly compensates for drag. If a difference exists between the driving force and drag, the body accelerates or decelerates, depending on the sign of this difference. In the case of an airplane, the force exerted by the propellers or the jet engines provides the propulsion. In the case of a bird, the efforts of the muscles beat their wings, which generate simultaneously lift and propulsion due to the release of a vortex with each beat. The kicks of a swimmer, with or without flippers, also release vortices in their wake; these are often visible on the surface. This is the force that propels the swimmer. A cyclist, even on a horizontal road and with no headwind, must pedal to overcome the resistance to his motion, which consists both of drag and of the mechanical friction linked to his bicycle. A javelin has no

---

9 Strictly speaking, the word *section* designates the cross section of a hull, with the *midship section* being the largest of all for a given ship. This term thus comes from the maritime vocabulary and we borrow it here to denote the equivalent quantity for an airplane wing.

10 NEWTON's second law of motion is more general than announced in the text above and holds only in an inertial reference frame. We have already invoked this law and we shall return to it to examine how to adapt it to relative motion; notably to motion within the frame of reference of Earth, which is permanently rotating. Insert 12.1 is devoted to this subject.

driving force once it is launched but advances by consuming its momentum, which diminishes monotonically along the trajectory so that the javelin slows until it sticks into the ground. The case of a glider is more original: the experienced pilot searches out ascending airflows, such as thermals, toward which he guides his glider to gain altitude. In this case, drag, which opposes the velocity, can have an upward component, which adds to the lift and participates in lifting the glider.

We have repeatedly noted that air is a fluid with very small viscosity,[11] about one thousand times less than water, and one hundred thousand times less than motor oil. The result is that friction drag is often much smaller than pressure drag, particularly in the case of poorly shaped objects. The big difference between the drag on a cyclist that must penetrate through air and that on a swimmer that must penetrate through water, about one thousand times denser, is essentially due to the different densities of these two fluids. We know that marine animals, notably dolphins, have good profiles despite a poor lift-to-drag ratio and possess a skin lubricated naturally by organic molecules that they themselves secrete. Birds are well aware of how to reduce their drag by tucking their feet into their plumage and by elongating themselves as much as possible in the direction of their velocity. The passage of a flock of pink flamencos above the swamps of the Camargue in France and the majestic soaring of eagles above the Alps illustrates well how these birds have developed their know-how, and even their anatomy, to exploit this means of minimizing their efforts.

2. Why are airplanes so noisy?

Let us investigate the origin of noise emitted by airplanes, helicopters, and wind turbines. This noise constitutes in fact one of the characteristics of our aerial environment and is often cited as a significant nuisance. It is often invoked to justify the installation of airports as far as possible from inhabited zones, and wind turbines at sea rather than on land. Just as guitar strings or the lash of a whip emit sound by shaking the air molecules that surround them, so do flying or rotating air foils or the exhaust jet of a jet engine. The intense noise produced by these objects results from the fact that they generate large unstable vortices, within which a depression

---

11 The dynamic viscosity of a fluid, often denoted by μ, is the viscosity relevant for evaluating friction and has units of poiseuille (1 Pl = 1 kg m^{-1} s^{-1}) in honor of the 19th century French doctor and physicist Jean-Louis Marie POISEUILLE. Under normal conditions, the dynamic viscosity of air is of the order of 1.8×10^{-5} Pl. In comparison, under the same conditions, that of water is of the order of 10^{-3} Pl and that of lubricating oil can reach 1 to 10 Pl. In studies of how viscosity affects flow, the kinematic viscosity of the fluid $\nu = \mu/\rho$ is often preferred (this is the ratio of the dynamic viscosity to the density). The kinematic viscosity has units of stokes, in honor of the British scientist in fluid mechanics George Gabriel STOKES who made significant contributions to the understanding of viscous friction. Note that the ratio between kinematic viscosities differ greatly from those between dynamic viscosities, to the point that the kinematic viscosity of air is 15 times that of water whereas its dynamic viscosity is 100 times less.

is created by the centrifugal force. Once they form, these vortices, which are vital for lift, move very rapidly with the airplane wing or the propeller blade. As a result, these unstable depressions shake the surrounding air and these perturbations propagate at the speed of sound (340 m s^{-1} near the ground). Except for exceptional cases, such as that of a supersonic airplane to which we shall return later, this speed is greater than the speed of the vortex that generated the perturbation. As a result, the noise can propagate in all directions, including upstream of the emitting source.

To make these concepts more precise, let us do a little exercise by imagining the very simplified but typical situation illustrated in figure 4.6.

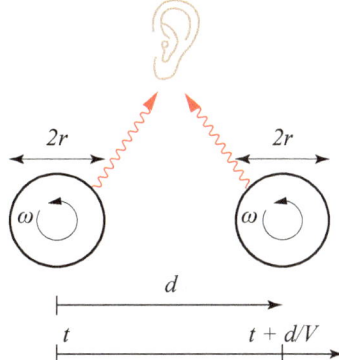

Figure 4.6 - A vortex moving at speed V occupies different positions at two instances separated by the time d/V. The central nucleus of radius r turns as one with angular speed ω. The vibrations generated in the surrounding air have the frequency V/d and the associated variations in pressure per unit length are of the order of $\rho\omega^2 r$.

A vortex chosen from a long sequence of identical vortices covers the distance d at speed V. We assume that, within the central nucleus of radius r, the air starts to rotate at angular speed ω. All nearby observers receive the vibrations generated in the air by the pressure fluctuations linked to the passage of each vortex, which require the time d/V to cover the distance d. The observer thus detects a vibration whose frequency is of the order of V/d. If we take, for example, $d = 0.1$ m and $V = 100$ m s^{-1}, the frequency of the noise emitted is of the order of 10^3 Hz, well within the audible range. To estimate the intensity of the sound emitted, we evaluate the fluctuations in pressure due to the centripetal depression within each vortex. This is at most of the order of $\rho\omega^2 r$, a value that could only be justified if the vortex rotated as a solid block. With $\rho \approx 1$ kg m^{-3}, $\omega \approx 50$ rad s^{-1} (or eight revolutions per second), and $r \approx d \approx 0.1$ m, we obtain a pressure fluctuation of the order of 250 Pa, which is 0.25 % of normal atmospheric pressure. Despite the relatively small value, this perturbation in the ambient pressure leads to audible noise of the order of 140 decibels,[12] which is roughly equivalent to a thunderclap, and

---

12　The bel was chosen to characterize the intensity of fluctuating physical phenomena, such as sound and many other physical quantities, in honor of the physicists Alexander Graham BELL, born in Scotland at the end of the 19th century and resident successively of Canada and the United States. By definition, the bel is the common logarithm of the ratio of the power of the phenomenon (P) to a reference power (P_0): $X_{bel} = \log_{10}(P/P_0)$. The decibel (dB) is one tenth of a bel. For sound, the power is proportional to the square of the amplitude of the pressure fluctuations, and a reference

much greater than the sound emitted by most household appliances (which ranges from 40 to 80 dB, where dB denotes decibel).

In reality, we have largely overestimated the intensity of the noise because we assumed that a vortex of radius r turns as one like a solid object with angular speed ω. This assumption of rotation as a solid is only justified for large vortices, which have, in effect, a rotating central nucleus, whereas the internal structure of smaller vortices, which are strongly sheared, is much more complex. As a result, the dimension of the central nucleus is much smaller than the size of the vortex. The vortex structures surrounding wings and propellers are effectively organized, as we have seen, to generate lift; but they are also very turbulent and, as such, depart strongly from the concept of rotation as a solid. This departure significantly reduces their centripetal depression and the noise generated as they pass by. To convince ourselves, assume that the angular speed is 10 times less than in the previous paragraph, or about 5 rad s–1. The centrifugal depression become 100 times weaker, its logarithm decreases by two, and the noise intensity is of the order of 100 dB (20 times 5 instead of 20 times 7).

The case of airplane propellers and the blades of helicopters or wind turbines merits an additional remark. Apart from the vortex released by each blade, which explains the lift transmitted to the axis, the ensemble of the propeller also forms a rotating structure of even larger scale. Even if we account for the fact that the air that traverses the propeller cannot rotate as one with it, an additional rather large vortex tends to form, roughly of the diameter of the propeller. This explains the loud noise created by helicopters, which is significantly louder than an airplane of similar power. Also noticeable is the choppy effect superimposed on the noise at the frequency of the passage of the blades in a given direction. These quantitative estimates show that the noise associated with a strong lift cannot be moderated because the lift results from the creation of fairly-well-structured and intense vortices that generate soundwaves as the travel.

Noise from jets exiting the nozzles of jet engines is explained in a similar fashion. As shown in figure 4.7, large spiral vortices form in the region where the jet separates from the nozzle wall and starts to mix with the external air; these vortices are generated by the KELVIN-HELMHOLTZ instability.[13] The first pair of vortices forms two successive toroids of the same sign, which tend to wrap one around the other while advancing rapidly in the wake of the jet engine. This relatively well-organized structure contributes most to the engine noise. The goal of specialists is thus to disorganize this structure to reduce this nuisance. For this, various methods may be used, such as attaching teeth to the edge of the exit nozzle to create

power 2×10^{-5} Pa is generally used. Given a pressure fluctuation of 250 Pa, the preceding formula yields $X_{Bel} = 2 \log_{10}(250/2\times10^{-5}) \approx 14$, or 140 dB.

13 Hermann Ludwig VON HELMHOLTZ was a German doctor and physicist of the 19th century, known notably for his work in the domain of acoustics.

three-dimensional perturbations that are expected to reduce the noise by breaking up the largest vortices. Needless to say, this solution butts against an inevitable penalty, because it also reduces the thrust of the jets against the nozzle; in other words, the propulsive force.

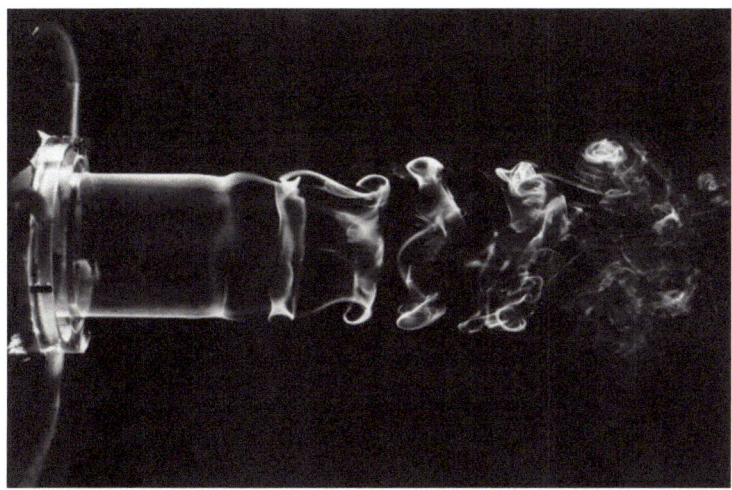

Figure 4.7 - View of a cylindrical jet of moderate REYNOLDS number (Re = 5600), showing the large, spiral, fairly-well-organized, and extremely noisy vortices that form within a few diameters of the source, and the small turbulent disordered structures that develop downstream, which are formed rather from filaments or sheared layers. The main source of noise in such a jet resides in the displacement of the large spiral vortices. [© KURIMA - KASAGI - HIRATA]

3. Shock wave and sound barrier

The mechanisms described at the end of the preceding section regarding noise from jet engines do not change if the airplane flies at supersonic speeds. The novelty lies in the fact that, when the speed of the airplane with respect to the ambient air exceeds the speed of sound, the sound waves, despite being emitted in all directions, cannot reach the front areas of the airplane because the airplane moves faster than they do. A sort of boundary thus exists between the regions that the sound waves can reach and those that are out of range. This boundary is called the *shockwave*. It is essentially fixed to the airplane, which carries it along.

Without hearing or seeing a shockwave, we shall try to understand it through a simplified analysis of the mechanisms that affect it. As an airplane travels a distance V per unit time, the soundwaves emitted travel a distance c. The soundwaves that move perpendicularly away from the trajectory of the airplane move the farthest from the airplane. Consequently, the soundwaves are limited to a cone around the airplane, as shown in figure 4.8. The half-angle θ of the cone is given by $\tan\theta = c/V$. Schematically, this boundary, which is called the *shockwave*, separates the upstream air, which knows nothing yet of the airplane because no soundwaves have traversed it, from the air inside the cone, through which the soundwaves emitted by the airplane have travelled. This boundary has a finite thickness of the order of a wavelength,[14] which is so thin with respect to the other length scales that it is often considered as a discontinuity (i.e., as if it had zero thickness).

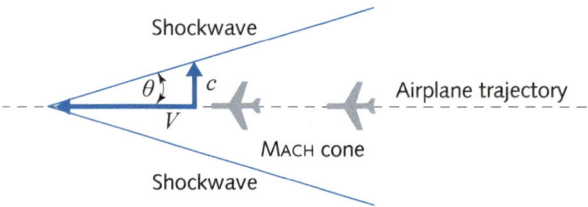

Figure 4.8 - Schematic representation of a shockwave, which defines the MACH cone around an airplane in supersonic flight ($Ma > 1$).

From the viewpoint of the essential physical mechanisms, retain simply that the shockwave is the region where all the waves previously emitted over a long period of time and whose energy has not yet been dissipated come together and superimpose over each other. The airplane has overtaken these waves because it travels faster than sound.

The term *shockwave* is justified by the analogy with the popping noise emitted when two objects collide at sufficient speed for the air between them to be evacuated at supersonic speeds, such as the hammer and anvil of a blacksmith. In this case, as for the explosion of the ionized channel of lightning (see chap. 3) or when the tip of a whip breaks the sound barrier, a significant quantity of acoustic energy is concentrated in a small volume and accompanied by a brief jump in pressure. Finally, in all these examples, the main fraction of sound energy emitted is dissipated in the shock wave. The ratio $Ma = V/c$ is known as the MACH *number*,[15] and the cone with half-angle θ is called the MACH *cone*.

In stable supersonic flight, an airplane is actually accompanied by two shockwaves: one starting at its nose and the leading edges of its wings, and the other starting at the leading edges of the tail assembly (fig. 4.9). We can consider that the air between the two shockwaves accompanies the airplane at its supersonic speed, except far from each side, and that the second shockwave restores the normal subsonic conditions behind the airplane. Upon hearing this double shockwave, the human ear cannot distinguish between the two closely spaced shocks but hears instead a violet boom, commonly called a *sonic boom*. The shock, which can be

---

14 The wavelength is the distance between two successive wave crests or troughs; in other words, the distance traveled in one period. For sound at a frequency of 10^3 Hz that moves at a speed of about 300 m s^{-1}, the wavelength is in the neighborhood of 30 cm.

15 The number $Ma = V/c$ is called the MACH *number* in honor of the discoveries on supersonic motion by the physicist Ernst MACH during the 19th century.

strong enough to break windows, is heard on the ground only under special conditions because, in peacetime, supersonic flights are forbidden above inhabited zones. Thus, during its transatlantic flights, the Concorde was authorized to break the sound barrier only above the ocean.

Figure 4.9 - Photograph of a U.S. Air Force F-22 Raptor in supersonic flight, taken from the aircraft carrier USS John C. STENNIS. The condensation of water forms clouds that delineate the depression zones. The large triangular cloud shows the main MACH cone, which forms starting at the leading edges of the wings. Note also the contrails where water vapor condenses. The water vapor comes from the jet engines, which are placed close to the wingtips.

Between the subsonic regime ($Ma < 1$), which necessarily exists after takeoff and before landing, and the supersonic regime ($Ma > 1$), which may be attained during the fastest part of the flight, is an intermediate regime called *transonic*. This transition was feared for many years by aviators, notably those that, during the Second World War, pushed their airplanes to the maximum. When they approached transonic conditions, they felt a significant hardening in the flight conditions, accompanied by a large increase in drag. The hardening is what brought these aviators to use the term *sound barrier*, which is not a physical reality comparable to the shockwave, but which represented nonetheless a significant threshold to cross[16] in the progress of aeronautics.

Conclusion

In the atmosphere, inert objects heavier than air, such as stones, fall toward the ground and lighter objects, such as warm smoke, rise higher. However, airplanes and helicopters, like birds and insects, have the capacity to fly because they are capable of simultaneously generating a lift that can overcome their weight and a driving force that can overcome their drag. The trajectories of these objects and animals, although sometimes curious, such as for balls and balloons, always result from the combination of these forces. The noise of propellers, the din of jets coming out of jet engines, and above all the sonic boom of supersonic airplanes, constitute a definite auditory nuisance that significant research efforts are aiming to diminish. Real progress is happening, but the noise represents such a small fraction of the energy required to make these vehicles fly that it has proven extremely difficult to reduce it without compromising the performance of the aircraft.

---

16 The first to break the sound barrier was the American pilot Charles YEAGER in 1947, aboard a Bell X-1 airplane.

Chapter 5

The tranquil sea

The sea is the vast reservoir of Nature.
The globe began with sea, so to speak; and who knows if it will not end with it?
In it is supreme tranquillity.

(Jules VERNE, *Twenty Thousand Leagues under the Sea*)

© Springer International Publishing AG 2017
R. Moreau, *Air and Water*,
DOI 10.1007/978-3-319-65215-3_5

L et us now engage in a fairly general overview of the seas and oceans, which we shall attempt to demystify by interjecting some anecdotes from their history—a history that is closely linked to that of Earth itself. As we did with the atmosphere, we start by examining the state of this immense mass of water, more or less salted depending on the particular region of the globe and sufficiently dense to prevent light from reaching its great depths, thereby retaining a part of their mystery. To observe these seas they must be navigated, which brings to our attention the remarkable stability of surface ships compared with submarines. We give a panorama synopsis of the gigantic thermohaline circulation,[1] whose best-known part is the Gulf Stream in the North Atlantic. Slowly but surely, with its power and inertia, this large marine loop drains the five oceans and constitutes the main current about which agitate multiple secondary currents that operate on the scales of the individual seas. We start by observing this global circulation and leave for the next chapter the fluctuations that ceaselessly perturb it throughout its course.

1. The sea at rest

1.1. A first panoramic glimpse

Seen from space, the surface of Earth appears mostly covered by the oceans, which occupy about 71% of the total surface area. The main characteristic of these large salted spaces comes in fact from the freshwater that is injected by rivers, streams, and precipitation and that escapes through evaporation, leaving behind to accumulate all dissolved and suspended matter. Since the formation of the oceans,[2] an equilibrium has been achieved between these various chemical inputs and their fixation in the sediments on the sea floor, so that the present chemical composition of ocean water, notably in terms of salt, is fairly stable. The oceans thus constitute a continuous domain where water, our other life-giving fluid after air, makes extremely long stopovers. This factor distinguishes the oceans from lakes, even the largest, which are traversed and drained by rivers or streams that rapidly renew their water.

During the slow evolution of our planet, plate tectonics completely reshuffled the number and positions of the continents and oceans. Recall simply that, during

---

1   The adjective *thermohaline* refers to the double mechanism that drives this circulation: one part is the temperature differences, the other is the salinity; both modify the water density. The suffix *haline* comes from the Greek words *alos* (ἁλός, salt) and *alinos* (ἅλινος, salin).

2   Earth formed 4.6 billion years ago; water in its liquid state only appeared much later, when the temperature dropped below 100 °C, and the first oceans formed later still.

the time of Pangaea, about 300 million years ago,[3] only a single large ocean encircled this lone continent. Today, according to the International Hydrographic Organization (IHO), three oceans exist. The largest is the Pacific, situated between Asia/Australia and the Americas. Its surface area accounts for roughly half of the total surface area of all the oceans and it alone covers a third of Earth's surface. Its predominance in surface area is no doubt the reason that its median meridian was chosen as the International Date Line. The Atlantic Ocean, located between the Americas and Europe/Africa, is the second largest ocean in terms of surface area and represents 30% of the total ocean surface area. It is much better supplied with fresh water because it receives the output of several large rivers, such as the Amazon, the Congo, and the Saint Lawrence. The Indian Ocean, the third in terms of surface area, represents about 20% of the total. It is almost entirely within the Southern Hemisphere, between Asia, Africa, and Australia.

Despite this official classification, our planet is commonly credited with five oceans rather than three by identifying the Southern Ocean, which envelops Antarctica, reaches up to about the 60th parallel, and accounts for about 6% of the total ocean surface area, and the Arctic Ocean in the north, which is bordered by Siberia, Scandinavia, Greenland, and North America and accounts for about 4% of the total ocean surface area. The Arctic Ocean is rather shallow and is partially covered by the Arctic Ice Sheet.

Within this scheme, the seas appear as rather small and individualized marine sub-domains. For most, the local geography justifies their individualization, as for the Mediterranean Sea, which is connected to the Atlantic Ocean only through the Straits of Gibraltar,[4] which are so narrow that the Atlantic tides pass almost unnoticed along the coasts of the Mediterranean. In contrast, the English Channel opens wide onto the Atlantic and is subject to strong tides, which are slowed between England and France by the Pas-de-Calais before they join up with those of the North Sea to be again slowed by the straits between Denmark and Sweden, after which they finally arrive almost imperceptibly at the Baltic Sea. In other cases, such as for the Caribbean Sea, even if the geographic separation is very real in the form

---

3   The supercontinent Pangaea dates from the Paleozoic geologic era, between 500 and 250 million years ago (mya). The Paleozoic era followed the Precambrian (−2500 to −500 mya) and preceded the Mesozoic (−250 to −65 mya), the Cenozoic period (−65 to −1.65 mya), and the current Quaternary period.

4   The Mediterranean Sea is the modern-day remnant of the ancient Tethys Ocean, which existed before the African and European plates came together, with the former diving under the latter and lifting it. This trajectory began over 100 mya and continues today, giving rise to violent earthquakes. It thrust up the Alps, formed the Pyrenees, and explains the formation of the Mediterranean volcanos. About 5 mya, the Strait of Gibraltar closed, isolating from the Atlantic a sort of large lake with ever increasing salinity and ever lowering water level. This lasted until the pressure difference between the oceanic face and the lakeside face caused the rupture of the rocky seawall between Gibraltar and Tangiers. A gigantic rush of water equalized the water levels and, in so doing, carved out the modern-day strait, which is about 14 km wide. This re-equilibration was fast, requiring only some 40 years.

of the long archipelago of islands stretching from the coasts of Venezuela to the peninsula of Florida, the individualism of the sea is more justified by the history of the coastal states and the successive conquests.

The study and protection of the seas and oceans are coordinated by the IHO, which was founded in 1921 under the initiative of ALBERT 1st, Prince of Monaco. This organization includes as members more than 80 states with maritime borders, and its headquarters are in the Principality of Monaco. One of its better-known activities is the creation of digital navigation maps, which are progressively replacing the traditional printed maps of questionable precision, being based only on successive observations by mariners.

Closed seas, such as the Caspian Sea, the Aral Sea, the Dead Sea, and the Black Sea, are special cases whose geological histories and recent developments are specific to each. The Caspian Sea receives the large flow of the Volga and recycles it by evaporation and infiltration into its sandy coast. It is 28 m below the average ocean level and has a low salinity. Although it currently is limited to the bottom of a rather small basin, in other times it formed an enormous sea together with the Black Sea, the Aral Sea, and the Arctic Ocean. The Aral Sea, which is in peril[5] of disappearing due to its supply rivers being diverted for agriculture over the course of the 20th century, is currently the subject of international aid efforts financed by the World Bank. The first of these efforts, which consisted of building a dam on the Syr-Darya River, was completed in 2005 and resulted in a 6 m increase in the level of the North Aral Sea. But the future of the South Aral Sea remains very uncertain. The Jordan River is also partially diverted for irrigation and supplies the Dead Sea, which has also lost a third of its surface area over the last fifty years and now holds the record for salinity. Its free surface is 417 m below the average level of the oceans. Canals have been envisaged to connect it to the Red Sea or to the Mediterranean, but these have yet to see the light of day.

The particular case of the Black Sea is very interesting, because its history after the last glacial period is the object of debate, or even controversy. The presence of freshwater sea shells within its current underwater sediment seems to establish that, at the end of the Würm glacial period, this sea was a freshwater lake, called the *Pontic Lake* in ancient texts such as the Bible. An isthmus several kilometers across separated the lake from the Sea of Marmara, which today connects directly with the Mediterranean Sea via the Dardanelles Strait. The water level in Pontic Lake was 180 m below the average sea level. The subsequent melting of the glaciers raised the level of the Mediterranean and the Sea of Marmara such that, according to American Geologists from Columbia University, William RYAN and Walter PITMAN, the valley of the Bosporus would have been flooded rather rapidly

---

5   The Aral Sea has lost three-quarters of its surface area since 1960, which is when the Karakum Canal began operation to irrigate the cotton fields of Uzbekistan and Kazakhstan. Since 1989, the North Aral Sea and the South Aral Sea have become separated bodies of water.

by salt water from the Sea of Marmara. This flood would have happened some 7000 years ago. A cascade with a flow of over 500000 m^3 s^{-1} would have spilled into Pontic Lake, thereby bringing the heavier salt water into the lighter freshwater and equilibrating the water levels. The lake shores would have receded by 1 km per day, chasing the agricultural population already present there toward Asia Minor and Mesopotamia. The mythical flood may have thus been born.

The relatively stable state of the Bosporus that we know today contains effectively two superposed currents: a deep current of salt water directed toward the Black Sea and a freshwater surface current directed toward the Sea of Marmara. According to W. RYAN and W. PITMAN, today's deep current would be the remains of the tremendous cascade of 7000 years ago. This theory is contested and two other scenarios are now proposed. One of them excludes the idea of such a rapid and catastrophic cascade and prefers a slower evolution; the other proposes a series of very-slowly-damped, very-long-period oscillations. The narrowness of the straits travelled by many ships and its strategic position, which results in the passage of many submarines, explains the accidents that occur there[6] and hinders any research that may alleviate the controversy.

1.2. Pressure, temperature, and salinity of seawater

As a first approximation of reality, imagine first that these immense aquatic spaces are in a state of absolute rest. The horizontal scales are of the same order of magnitude as the radius of Earth (thousands of kilometers) and the vertical scales can exceed 10 km. These quantities are remarkably similar to those of the atmosphere; however, the thickness of the troposphere varies little and very slowly from one point to the next, whereas the depth of the seas varies significantly between very shallow zones in lagoons and the great depths in the deepest oceanic trenches.

One of the first questions to ask ourselves concerns the variations in the density of this water, which is one of the most important properties characterizing the state of matter in equilibrium. Recall first that, under standard conditions (normal pressure of 1013 hPa, a temperature of 3.98 °C, zero salinity) the density of water is exactly 10^3 kg m^{-3}. We may remember that these conditions gave us the definition of the kilogram, which is the maximum mass of a liter of freshwater.[7] The density of seawater varies particularly with temperature and salinity and much less with

---

6   In 1994, a collision between two supertankers, followed by an enormous fire, caused significant pollution. The last large shipwreck in the Bosporus, dating from 2005, was that of a Panamanian cargo ship transporting metals.

7   Since 1889, by a decision of the International Bureau of Weights and Measures (IBWM), the kilogram is defined as the mass of a platinum-iridium prototype, which is conserved in Breteuil, France. The kilogram is one of the fundamental units of the International System of Units (SI) and is the last to be based on a man-made object; in other words, on an artifact. Considerations are underway to redefine the kilogram.

pressure, which often leads us to consider this fluid to be incompressible.[8] The salinity, mass fraction of salt, or again the mass of salt per unit volume of seawater per mass of this volume, is normally counted in parts per thousand (ppt), but we shall use the units of g kg^{-1} to ensure clarity. In lakes and streams, the salinity is almost zero, or rarely exceeds several g kg^{-1}. Although its average value in the oceans is 35 g kg^{-1}, it can attain and sometimes exceed 50 g kg^{-1}. In the Black Sea, it is 12 g kg^{-1}. In the Dead Sea, its extremely high value of 275 g kg^{-1} ensures the absence of practically all life forms, be they animal or plant.

In the seas, the water temperature varies between 2 and 30 °C, even if figure 5.1, which relates to the Mediterranean Sea, suggests a more limited range of between 13 and 24 °C. The variation in density, displayed in figure 5.2, shows a maximum around 3.98 °C for freshwater and decreases from 13 to 30 °C. Around 15 °C, the slope of this curve is rather constant, justifying the linear BOUSSINESQ approximation,[9] which is very popular for studying natural convection. Moreover, during summer, the surface water of the warmest seas can reach 26 to 30 °C, which often leads to cyclones, as we saw in chapter 2.

The highest layers of water (i.e., the surface water) are warmed by solar radiation and subject to constant thermal exchange by conduction and convection with the atmosphere. Mixing by the waves and turbulence manages to homogenize the temperature within the first tens of meters (between 0 and −50 m). In contrast, at great depths (below −120 m), any exchange is limited purely to conduction and becomes notably weaker, to the point that approximating these depths to be at rest is well justified, even if deep, extremely slow currents couple with the surface currents to weakly amplify the apparent thermal conductivity of the deep waters. Between these two zones, common practice is to identify a thin layer (between −50 and −120 m) called the *thermocline*, where the temperature drops about ten degrees Celsius. This drop varies strongly with the seasons because the surface temperature changes whereas the temperature of the deep water does not vary. Note from figure 5.1 that the thermocline in the Mediterranean Sea builds up in springtime in keeping pace with the increase in sunlight. The maximum temperature difference of about 10 °C is attained in August, and it diminishes over autumn

---

8   To consider the propagation of sound in seawater, as in freshwater, this approximation must be rescinded to account for the compressibility of this fluid.

9   BOUSSINESQ's formula for water is an equation of state that is linearized about a reference point (T_0 = 288 K, p_0 = 1013 hPa, S_0 = 0, where ρ_0 = 1000 kg m^{-3}), which amounts to replacing the real curve by its local tangent. Let β_T be the coefficient of volume expansivity, β_p be the coefficient of volume compressibility, and β_S be the coefficient of variation due to salinity. With these definitions, we form the following expression: $(\rho - \rho_0)/\rho_0 = \beta_T(T - T_0) + \beta_p(p - p_0) + \beta_S S$. The estimated coefficients are $\beta_T = 2 \times 10^{-4}$ K^{-1}, $\beta_p = 4 \times 10^{-5}$ hPa^{-1}, and $\beta_S = 8 \times 10^{-4}$ kg g^{-1}. This expression gives a satisfactory approximation of the density of water for most applications, except near the maximum density of ρ = 1000 kg m^{-3} for T = 3.98 °C, from which the kilogram was initially defined. It is remarkable that the effect of compressibility is about tenfold less than the two other effects, which justifies the frequently used hypothesis of incompressibility.

and the beginning of winter. In February and March, the thermocline disappears, making the temperature essentially constant at 13.5 °C throughout the depths. It is noteworthy that, above and below the thermocline, the temperature as a function of depth continues to increase, although at a very low rate. As a result, this deep layer heated from above is not at all subject to the RAYLEIGH-BÉNARD convective instability (see appendix).

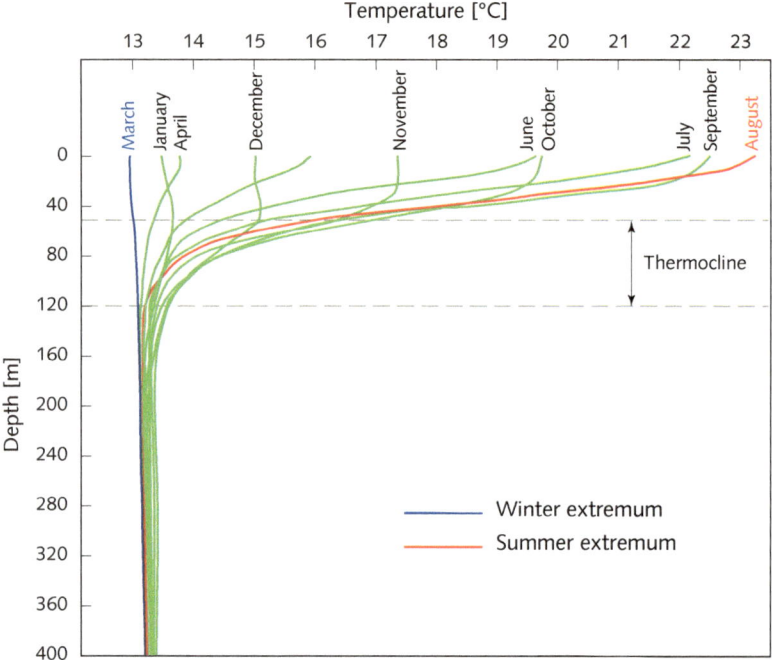

Figure 5.1 - Temperature distribution in the Mediterranean as a function of depth, showing the thermocline and its seasonal variations. [taken from LEVITUS *et al.*,1994 - World Ocean Atlas, NOAA]

The chemical composition of seawater is not simple. Dissolved in the water itself, which constitutes 96.5% of the total mass, we find a large fraction of the elements in the form of a complex mixture of anions, cations, and molecules. The ions are more than just the chlorine anion Cl^- or the sodium cation Na^+, even if these two largely predominant elements constitute the basis of marine salt. Table 5.1 lists in order of decreasing concentration the five main anions and cations of typical seawater, whose salinity is 35 g kg^{-1}. We find that the ratio between the concentrations of all these ions varies little from one sea to another,[10] so that, if we measure the concentration in one, we can deduce the global salinity.

---

10 This empirical property is known as DITTMAR's law in honor of the Scottish chemist William DITTMAR (1859–1951), who arrived at this law based on observations made during the Challenger oceanographic expedition from 1872 to 1876.

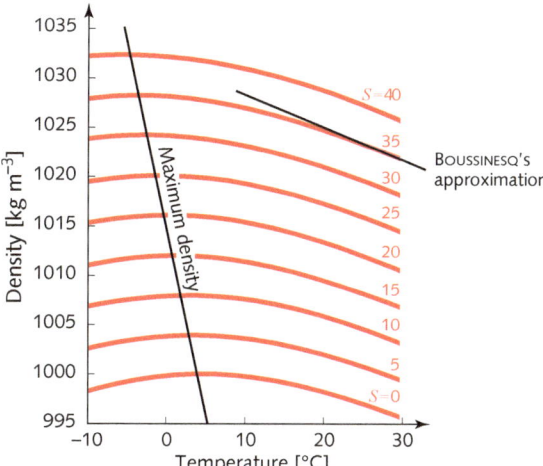

Figure 5.2 - Variation in seawater density as a function of temperature for various salinities S. Note the positions of the maximum densities.

These elements have various origins. Some ions come from the dissolution of continental rocks by rivers, which then transport these elements to the oceans where they remain for an extremely long time, with their concentration being increased by evaporation. A significant fraction of Na^+ cations appear to come from the initial ocean floor. And the origin of chlorine ions Cl^-, which are soluble in water, is often attributed to the outgassing of hydrogen chloride from volcanos.

Table 5.1 - Concentrations of main ions dissolved in seawater

Anions	g kg^{-1}	Cations	g kg^{-1}
Chlorine	19.352	Sodium	10.784
Sulfate	2.712	Magnesium	1.284
Hydrogen carbonate	0.108	Calcium	0.412
Bromide	0.067	Potassium	0.399
Carbonates	0.015	Strontium	0.008

Other than water and salts, we also find small concentrations of diverse molecules, such as boric acid $B(OH)_3$ (0.0198 g kg^{-1}) and carbon dioxide CO_2 (0.0004 g kg^{-1}), as well as nitrogen and oxygen. Remarkably, the concentration of carbon dioxide is much greater in the water than in the air (about 60 times more), although this is not the upper limit of the oceans' capacity to stock this molecule. This touches on a question currently under intense debate: the possibility of sequestering carbon dioxide in the oceans in order to decrease the concentration of this greenhouse gas in the atmosphere. Some suggest capturing this gas near the sources and injecting it deep into the oceans, despite the uncertainties regarding the potential reactions such as sensitive modifications of the pH [11] of seawater or the carbon cycle, whose effects remain very difficult to qualitatively evaluate at the time these lines were written.

Figure 5.2 also shows that, in salt water, the maximum density shifts to lower temperatures, less than 0 °C. Under ice floes, where the temperature ranges from 0 to 4 °C, certain phenomena, such as the morphology of the water-ice interface and the instantaneous variation of the frozen mass, cannot be explained without abandoning the BOUSSINESQ approximation and considering instead the real, non-monotonic variation of the density as a function of temperature.

1.3. The sea is neither flat nor round

Given that mass is not uniformly distributed within the terrestrial mantle, gravity cannot be uniform on the surface of Earth. This effect alone implies that the altitude of the oceans that are supposedly immobile must vary inversely with the local gravity; being a maximum where gravity is minimum, and vice versa, so that the product is constant. [12] As a result, the average free surface of the oceans cannot coincide exactly with the ellipsoid that we imagined at the beginning of chapter 1 where we likened Earth to a sphere slightly flattened at the poles. However, because all altitudes are measured with respect to sea level, we must specify how we arrive at this level, [13] or at least its average value. This requirement led oceanographers and geophysicists to choose the equipotential surface of the gravitational field that coincides best with the average level of the oceans. This rather peculiar surface, called the *geoid*, [14] is displayed in figure 5.3 and shows the extent to which the real ocean surface departs from the ellipsoid corresponding to the case of uniform gravity. This bumpy surface represents the variations in gravity on Earth's surface, both over the continents and over the oceans.

---

11 The acronym pH stands for *potential of hydrogen*. Its value represents the chemical activity of hydrogen ions (H^+) in solution and characterizes the acidity or basicity of the solution. Thus, in an aqueous medium at 25 °C, a solution is acidic if the pH is less than 7, neutral if the pH is 7, and basic if the pH exceeds 7.

12 In all flowing fluid media, the equation of hydrostatic equilibrium is $p + \rho gz = const$. Given that atmospheric pressure is practically constant at the surface of the oceans, their altitude at all points M, the average gravity g_0, and the average altitude z_0, could be deduced, were the oceans at rest, from the local gravity g_M by using $g_M z_M = g_0 z_0$.

13 Knowledge of the altitude is important in numerous domains. An example is hydrology, which must be based on very precise data to predict river flows and eventual floods following heavy precipitation. Another domain is the positioning and guiding of all vehicles, ships, airplanes, rockets, and satellites by the worldwide global positioning system (GPS).

14 Gravity, or weight per unit mass, comes from a potential. As for all vector magnitudes that have this property, it can be represented by a family of equipotential surfaces to which it is orthogonal. These surfaces are tighter, like lines on a topographic map, where gravity is maximal and farther apart where gravity is minimal. Schematically, we can consider the bumps of the geoid to correspond to the gravity minima, and the hollows to the maxima.

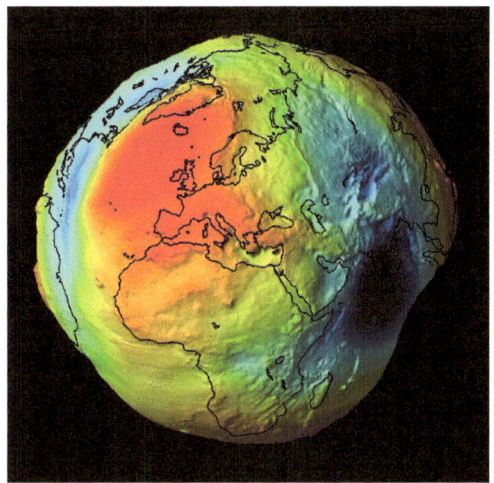

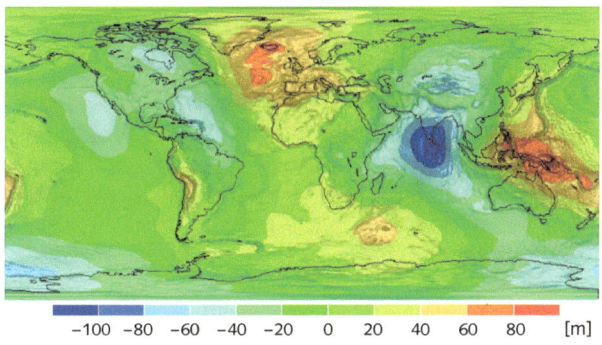

Figure 5.3 - The upper panel shows the geoid, which is the gravitational equipotential surface that best coincides with the average sea level [© ESA/GOCE High Level]. The lower panel shows a plane projection of the geoid [© ESA]. The extreme levels are visible (+80 m in southeast Greenland, −80 m in southern India), as are the levels relative to the Mediterranean Sea (+40 m near the Balearic Islands, −40 m to the southeast of Crete). These views were acquired by the GOCE satellite, which was launched by the European Space Agency in March, 2009 and dedicated specifically to the study of the gravitational field and the ocean circulation.

To know the average sea level over the entire globe, several long-term-observation programs have been undertaken. The first satellite dedicated to this mission was Topex-Poseidon, which was placed into orbit on August 10, 1992 by the French National Center for Space Studies (CNES) and stopped in January, 2006. In December, 2001, the satellite Jason was launched into a nearby orbit so that, until January, 2006, the observations benefited from crosschecking with complementary data gathered by the two satellites. The effort continues today with the Jason program and the recent Gravity field and steady-state Ocean Circulation Explorer (GOCE) program launched by the European Space Agency (ESA) in March, 2009. For nearly 20 years, these three satellites have thus provided precise measurements of the instantaneous levels of the seas and oceans. After filtering out all the periodic phenomena (tides, swells, and all sorts of waves), we now have a good knowledge of the average topography of the geoid.

Their average free surface is shown in figure 5.3, where the highest levels are to the southeast of Greenland (80 m above the average sea level) and the lowest levels are to the south of the Indian peninsula (about 80 m below the average sea level). In the particular case of the Mediterranean, the highest level (40 m above the average level) is to the west, near the Balearic Islands, whereas the lowest level (40 m below the average level) is to the southeast of Crete, in the eastern basin. Remarkable is the fact that the Mediterranean is composed of two geological basins: a relatively shallow western basin and an extremely deep eastern basin, separated by a zone of high shoals between Sicily and Tunisia. The average depth is 1500 m, with the deepest abyss of over 5000 m being located off the coast of Greece. In comparison, the English Channel is much less deep (an maximum depth of 172 m).

We will see later that the ocean currents are blocked and reoriented by the continental coasts, which thus play the role of a wall or deflector. This reorientation implies the existence of zones of above-normal and below-normal pressure (i.e., overpressure and underpressure), which must be connected to variations in the level of the free surface of the seas. However, this effect can only constitute a small correction, on the order of one meter, that is relatively localized within coastal regions; we can therefore neglect it with respect to the influence of gravitational variations.

Some numbers are useful here to better understand the magnitude of the pressure variations as a function of the ocean depths. Recall that traversing a 10-m-thick aqueous layer suffices to generate a pressure difference of the same order of magnitude as atmospheric pressure, which itself is due to the total weight of the atmosphere—a layer some 10 km thick. This is completely consistent with the ratio of the densities: 10 m of water is approximately equivalent to 10 km of air. These numbers allow divers to understand why their eardrums hurt: they are subjected to strong variations in pressure. At a depth of 400 m, which is often attained by modern submarines, the pressure exerted by the water on the hull is on the order of 40 bars, so the interior pressure, which is maintained near atmospheric pressure for the benefit of the crew (1 bar or 10^3 hPa), is far from balancing it. We can understand now why these vessels are made from thick sheets of metal, capable of withstanding this difference in pressure between the two faces. In comparison, airplanes may be constructed from lighter alloys[15] based on aluminum and titanium because they need only withstand the difference between the interior pressure, maintained at about 1 bar, and the external pressure, which is always between 0 and 1 bar. The lightness of these materials is also required to facilitate

---

15 The main fraction of the mass of a modern airliner (about 80%) is made of alloys such as *TA6V*, which consists of titanium, aluminum (6%), and vanadium (4%). This represents a good compromise between lightness and mechanical resistance.

flying. Bathyscaphes, which are submersibles designed to dive to the deepest abysses of the ocean, must withstand enormous compressions[16] (500 bars at a depth of 5000 m), even if the interior pressure can reach quite high levels when they are not occupied.

2. Sound and light in seawater

Sound and light propagate through water and allow us to observe from a distance not only diverse objects but also the topography of the sea floor. Given the distribution of temperature, salinity, and pressure, one can deduce the speed of sound at all points, which is on the order of 1500 m s^{-1}, much higher than in air, although it varies somewhat with depth. As a result, if we measure these distributions, we can know the speed of sound at all points, which then allows us to calculate the trajectories of sound waves. This property led to the development of sonar (which stands for SOund Navigation And Ranging). An emitter-receiver taken onboard a ship or on a diving vessel, or fixed to the bottom of the sea, emits sound at the appropriate frequency and captures the echoes. By analyzing the delay between the sonic emission and the reception of the echo, the sound trajectories are calculated by onboard computers or by those at a dedicated listening station on shore, allowing the escorts, helicopters, or frigates of naval forces to detect submarines, and for submarines to detect surface vessels. The same technique allows geographers to analyze the undersea topography and for fishermen to detect schools of fish.

In practice, to observe the extreme depths of the oceans, the vertical temperature and salinity profiles must be measured by submerging for a brief time ballasted objects equipped with captors that simultaneously measure these two quantities as a function of depth. These data are collected by onboard computers and then input into a numerical model to determine the speed of sound at all depths and to localize the origin of any echoes received. As an example, figure 5.4 shows a typical result for the speed of sound as a function of depth. Under certain conditions, the sound waves emitted from the surface are reflected from the thermocline and cannot reach the lowest layers. In this case, surface ships or helicopters must submerge equipment carrying emitters and receivers underneath the thermocline in order to explore the great depths.

Of course, the ocean floor, marine mammals, and schools of fish all create echoes, requiring particular procedures to recognize the detected object. Apparently, very sensitive human ears are still capable of beating computers in this delicate game, which consists of guessing the nature of the object that created the echo. To detect

---

16 The record for the deepest dive was made in the Mariana Trench in the northwest Pacific Ocean in March, 2012, by the American moviemaker James CAMERON in a specially designed bathyscaphe. He attained a depth of approximately 11 000 m.

schools of fish, fishermen, be they professional or amateur, use sonar with rather high frequency (several tens of hertz), so that the sound is rapidly absorbed, limiting the range to only moderate depths. In exchange, they can obtain actual images of the objects detected. The lowest frequencies, between 3 and 20 hertz, are used by the military to detect distant objects and by geographers to characterize with precision the undersea topography.

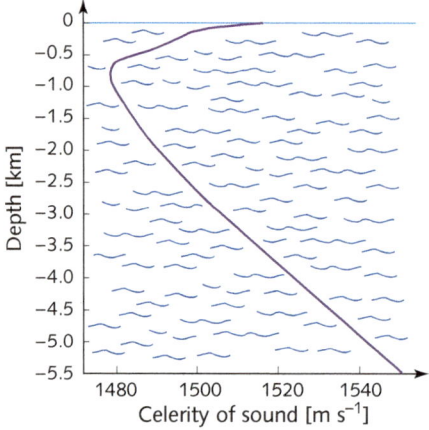

Figure 5.4 - Celerity of sound as a function of depth in the Pacific Ocean to the north of Hawaii. Data are from the 2005 World Ocean Atlas. The celerity of sound goes through a minimum at a depth of about 750 m.

The propagation of light in seawater is above all distinguished by a strong, wavelength-dependent attenuation. Even pure water, being simply eight hundred times denser than air, rapidly absorbs the energy of photons that, having come from the Sun, have traversed the essentially transparent atmosphere. And this absorption varies strongly with wavelength; in other words, with color. Infrared rays are absorbed within the first meter and warm this superficial layer, as we have seen above. Reds and yellows do not penetrate more than about ten meters. Only blues and greens reach greater depths but, all along their path, are subject to scattering, whereby water molecules and solute particles send off in all directions the intercepted photons. The composition of seawater, notably its salinity, only weakly affects the attenuation of light, which is very nearly that of distilled water. In contrast, suspensions, which generally contain particles much larger than water molecules, can strongly affect the scattering. This case is particularly relevant in cold seas, which contain more phytoplankton. These absorb more strongly in the blue than in the green, thereby shifting toward the green the color range of these seas rich in phytoplankton, and therefore in fish.

To conclude this discussion of the attenuation of light in seawater, we add only that, beyond several hundred meters, the marine environment is plunged in complete darkness. Divers only venture there when equipped with powerful lights. Undersea vessels are always equipped with sonar and their lights are turned on the moment the sonar detects a nearby seabed.

3. The remarkable stability of ships

To observe and know the ocean requires navigating on its surface and in its depths. We discover then that, even when it seems calm, the water is ceaselessly agitated, both by currents of moderate speed and by waves that vary greatly in amplitude. The next chapter provides the occasion to become more familiar with these effects. But, before engaging in this next step, let us try to understand the mechanisms behind the oscillations, which occur even during good weather, of surface ships and undersea vessels and extract the conditions required for their stability. This could help some to confront and accept both the pitching of boats, to which one becomes quickly accustomed (excepting perhaps when in a raging sea), and the soft rolling that is less worrisome. The fact that the stability is greater with respect to pitching than to rolling is simple to understand: when the prow of the boat is lifted by a wave, the length of the boat is sufficiently long for the righting moment, which is proportional to the lever arm, to restore the equilibrium position. In reality, an oscillation occurs about an average position, but the fact that the length of a boat is much greater than the width makes pitching of little worry. In contrast, with regards to rolling, we cannot help asking ourselves, notably if one has experience sailing small dinghies and has had the occasion to heel over on a first outing, what end is served by learning to right a boat? The fact is that boats capsize very infrequently,[17] even in strong seas. To what is due, then, their remarkably stability?

To simplify, we limit ourselves to considering only the roll and we start by imagining the case of a submarine to eliminate the free surface. A submarine is a sort of tube whose cross section is shaped like an ellipse, as shown in figure 5.5. The buoyancy force, which equals the weight of the displaced water, is applied at the center of gravity of the volume C, which is called the *center of buoyancy*. The mass of the submarine also has a center of gravity[18] G where the weight is applied and whose position depends on the arrangement of the masses that comprise the vessel. The position of the center of gravity may thus vary, notably when the crew moves about. At equilibrium, points C and G are aligned along the same vertical line. Imagine now a small tilt due to a roll. We immediately see from figure 5.5 that, if G is below C, the righting moment formed by the two nonaligned but parallel forces tends to bring the submarine back to equilibrium. In the opposite case, where G is above C, the righting moment tends on the contrary to accentuate the tilt. The

---

17 The naval schooner *Vasa* can be visited in Stockholm. This ship is named after the reigning Swedish family before BERNADOTTE and was built without respecting the condition for stability. She capsized after sailing several hundred meters on the day of her inauguration in 1628, before even attaining the sea. This ship was naturally conserved initially in the low-salinity waters of the Baltic Sea before being refloated in 1961, partially restored, and exposed in a museum specially designed for her conservation.

18 The center of gravity of a volume is its center of mass if gravity is uniform throughout the volume. The center of gravity of a planar surface is defined in the same way, but in two dimensions.

condition for stability of submersibles is thus clear: the diverse masses that comprise it must be distributed so that G is below C. This is a relatively severe condition.

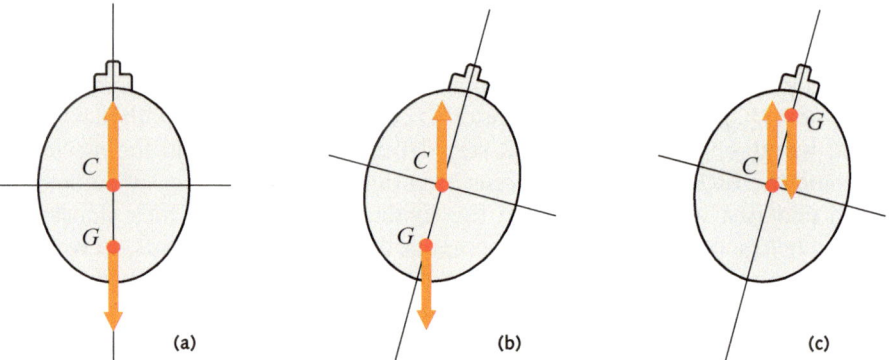

Figure 5.5 - Equilibrium and stability of a submarine. The arrows represent the weight applied at the center of gravity G and the buoyancy force applied at the center of buoyancy C: **(a)** equilibrium position, **(b)** righting moment when G is below C, **(c)** capsizing moment when G is above C.

The situation differs for a surface ship immerged in an ensemble of two fluids: water and air. One may think that, because air is 800 times less dense than water, it can be neglected. What is true is that the buoyancy due to air is negligible with respect to that of water. But it would be inexact to totally ignore air: it intervenes via the existence of the free surface, which applies rather strong forces to both the port and starboard sides. The submerged part of the ship is delimited by the hull and by the portion of free surface deleted by the presence of the ship, which is called the *waterplane area* and whose contour is called the *float line*. Together, the hull and the waterplane area delimit the submerged volume $\mathcal{V}$.

The novelty with respect to a submarine arises because, for a given submerged volume and weight, the width of the waterplane area can vary. For a relatively thin vessel, such as a long barge, all the weight is close to the axis as the vessel heels so that righting moment is small because all the forces due to water pressure are applied close to this axis. In contrast, for a wide waterplane area, such as shown in figure 5.6, the righting moment can be quite large because the hull makes contact with the free surface rather far from the axis. In other words, the righting moment benefits from a large lever arm.

This creates a point called the *metacenter* M that has no equivalence in submarines and which lies a distance $I/\mathcal{V}$ above the center of buoyancy C on the line defined by the buoyancy force (i.e., vertical). In this expression, I denotes the moment of inertia of the waterplane area [19] about the axis of symmetry and $\mathcal{V}$ is the submerged

---

19 All planar surface areas have two principle orthogonal axes that go through their center of gravity. For the waterplane area of a ship, these are the axis of symmetry that goes from the bow to the

volume. When the vessel is subject to small tilts, the metacenter remains fixed and the point C moves along a circular arc centered on M and with radius $I/\mathcal{V}$. The criterion for stability is that G remain below M rather than below C. The difference with respect to the situation with a submarine is significant: a ship can remain stable when G is above C provided that the metacenter is sufficiently high. All else being equal, this requirement demands a sufficiently wide waterplane area. In similar conditions, a submarine would be unstable.

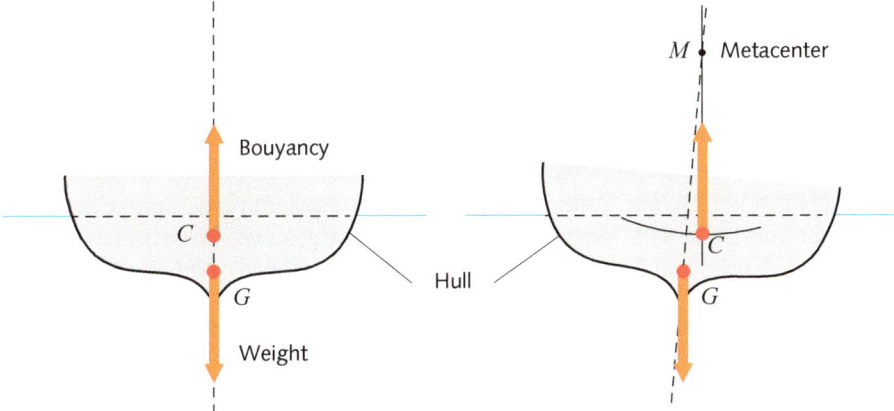

Figure 5.6 - A surface vessel is in stable equilibrium when the metacenter M lies above the center of gravity G. For small tilts, the center of buoyancy C shifts along the circular arc of radius $I/\mathcal{V}$ centered at M, which lies at the intersection of a vertical line through C and the vertical axis of symmetry of the vessel passing through G.

One way to obtain a large metacenter radius $I/\mathcal{V}$ to ensure the stability of the vessel consists of using a two hulls, both of which are displaced from the axis of symmetry. With the waterplane area being far from the axis, the metacentric radius becomes quite large. This idea led to the conception of catamarans. The trimaran has a main hull centered on the axis of symmetry and of sufficient size to accommodate not only the equipment necessary for navigation (sails, masts, ropes) but also the crew. Rather far out on each side of the main hull are two outrigger hulls, which are much lighter than the main hull. At equilibrium, only the main hull is partially submerged, but a modest heel suffices to bring a one of the two lateral hulls in contact with the water, which provides a lateral buoyancy force that, thanks to the long lever arm, adds a strong righting moment. The trimaran thus represents an intermediate scheme between a catamaran and a single-hulled vessel and offers the advantage of being very stable under strong tilts.

---

stern and the perpendicular axis. The general expression for the moment of inertia about each axis involves an integral. In the special case of a rectangle with a long side L and a short side l, the moment of inertia about the long axis is $Ll^3/12$. For a surface vessel, if the volume of the hull can be approximated as Llh, where h is proportional to the draught, the metacentric radius is given by $CM = l^2/12h$.

4. Global circulation in the oceans

This continuous, dense, and fluid medium that is the oceans is traversed by an organized set of quasipermanent currents that have a considerable effect on the climate and weather forecasts because of the enormous heat capacity and inertia of their immense mass. As we have already seen, the driving force behind the atmospheric airflows is the direct heating of seawater by the Sun, which is concentrated in the tropical regions where it leads to the high temperatures of the water above the thermocline. In return, the winds thus created sweep the surface water along via friction, thereby creating the slow but powerful oceanic circulation. Note that the kinetic energy of these fluid media comes entirely from the Sun. However, an important difference separates the aerial circulation from that of the oceans: the ocean currents are forced to contour around the coasts and the underwater ridges, which form impassable obstacles. Nothing comparable diverts the major airflows such as the jet stream. Moreover, as surface water cools or becomes saltier, it becomes denser, and these variations in density are capable of driving these waters deep into the ocean depths, giving the ocean circulation a three-dimensional character. In this new section, we shall attempt to understand this great thermohaline circulation, whose organization is remarkably stable on the scale of the five oceans.

The trade winds generate a westward airflow at the equator, which sweeps the topmost layer of water toward the west (see chap. 3). Whereas in the atmosphere, this wind from the east circles the globe without meeting any impassable barriers, the ocean waters pushed westward are diverted by the continents. Let us start in the mid-Atlantic, where the water is swept toward the coasts of North America, with a systematic deviation to the right (i.e., north) due to the CORIOLIS force. The light, warm surface waters depart from Cape Verde and head toward Florida, which blocks further progress toward the west. This obstacle stops the marine current and generates an overpressure, which is accompanied by an elevation in sea level that pushes the current back toward the east, while conserving its northward momentum. We are following here the well-characterized thermohaline circulation called the *Gulf Stream*, which is labeled 1 in figure 5.7.

The overpressure thus formed along the coast of North America pushes the Gulf Stream toward Europe at latitudes where the trade winds can no longer oppose it. The water, still rather warm and light, approach Spain, France, and Great Britain (see number 2) and warm the accompanying layers of air that also flow generally eastward, although at much higher speeds. This dominant wind from the west, which can be rather strong and unstable, is induced by the jet stream, which circulates at high altitudes. These two currents, one marine and the other aerial,[20]

---

20 Apparently the contribution of the Gulf Stream to the temperature difference between Western Europe and Eastern Canada, which are of the order of 10 to 15 °C at a given latitude, is much greater than that of the eastward airflow induced by the jet stream.

promote relatively mild temperatures in Western Europe, which are appreciated, especially in winter, in comparison with the temperatures in North America at the same latitude. A little farther along, the coasts of Northern Europe present another almost-impassable barrier[21] and direct the waters of the Gulf Stream westward, but still with a northward component, so that the marine current heads off toward Iceland and Greenland (number 3).

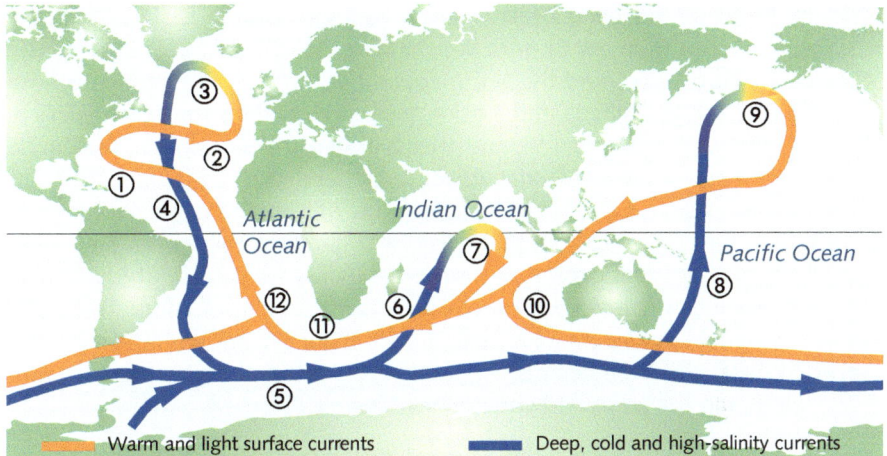

Figure 5.7 - Global thermohaline circulation through the oceans. The orange indicates the surface currents, and the blue indicates the deep currents. The numbers label the various sections discussed in the main text.

All along this first stage in the North Atlantic, from the Equator toward Greenland, evaporation progressively increases the salinity. In addition, the shift in latitude is accompanied by a cooling. Together, these two effects contribute to increasing the density of these waters. The ice floe then adds another important contribution: by injecting salt, which is less soluble in ice than in water, it further densifies the water, to the point of causing it to dive down to the depths. The surface current thus disappears but is prolonged by a cold, dense, and deep current (number 4), which can only head south because of the shallowness of the Arctic Ocean, all the while being subject to the CORIOLIS force. It is this force that, in the Southern Hemisphere, turns this current eastward toward Africa and around its tip instead of westward around the tip of South America. The next obstacle on this southward journey is Antarctica, which obliges this current to circumnavigate it from west to east (number 5).

Farther downstream, as seen in figure 5.7, a fork occurs in each new ocean, so that the flow from the Atlantic is shared three ways. Apart from the deep current that circumnavigates Antarctica, known as the *West Wind Drift* or the *Antarctic*

---

21 The Strait of Gibraltar is much too narrow to bleed off any significant fraction of the current. In contrast, the English Channel, on its scale, accepts a significant flow, which generates the large tides on the Normandy coasts. All the same, on the scale of the Atlantic Ocean, this flow remains modest.

Circumpolar Current, a first branch (called the *Mozambique current*) takes part of the flow toward Madagascar (number 6) and then toward the Indian peninsula (number 7), yet another impassable barrier. This flow climbs back up toward the free surface as it progressively warms and has no other option than to head off again to the south, deviating to the right (i.e., west) as it goes since it has reached the Northern Hemisphere.

The third part of this deep ocean current contours around Eastern Australia in the Pacific Ocean (number 8), forming the loops known as the *South Equatorial Current* (or *Kuro Shivo*) and the *North Equatorial Current*. They also encounter a barrier in the form of Siberia and Alaska, separated by the Bering Strait. This flow climbs to the surface (number 9), and orients itself southward with a western deviation, warming as it goes. When it arrives in the Southern Hemisphere, the sign of the CORIOLIS force changes, which reroutes some of its flow eastward (number 10), constituting the surface current that encircles Antarctica. The remaining flow joins up with the current that circulated around the Indian Ocean, and these combined currents (number 11) head west around the southern tip of Africa. This second part of the global circulation joins up in the Southern Atlantic with the current that encircled Antarctica (number 12), where they are swept along by friction with the trade winds and the westward equatorial airflow.

Another effect, due to the fact that Earth orbits the Sun, adds a non-negligible contribution to the thermohaline circulation, notably in the equatorial regions. The circle defined by the exterior of the torus swept out by Earth is longer than the interior circle, with the difference being about 80 000 km. By conservation of momentum of the overall system, this difference in distance traveled per year implies a difference in speed of about 220 km/day, or about 9 km h^{-1}. This effect drives an additional equatorial current, which adds to the impetus provided by the trade winds because it acts in the direction opposite that of the rotation of Earth about its axis.

We have just followed, step by step, this immense thermohaline circulation, which must be considered as the main oceanic circulation because of its gigantic inertia, its climatic consequences, and its own influence on the secondary oceanic currents. Some numbers are useful here to understand its primal importance. One of the most significant is no doubt the round-trip time for the circuit, which is of the order of 1600 years. The highest speeds measured are of the order of 10 km h^{-1}. But this number largely overestimates the average speed, which is doubtless below 1 km h^{-1} because of the numerous hindrances connected with the formation of enormous spiral vortices between the main current and the large fluid domains that it circumnavigates. As an example, figure 5.8 shows that the main current in the North Atlantic, between Florida and the European coasts, can be viewed as an enormous river about 100 km wide inserted into the ocean itself between immense decelerating vortices. Assume that this relatively warm branch of the Gulf Stream remains above the thermocline, so that its depth can be estimated to be about

50 m. From these numbers, we can immediately estimate the flow transported by this North Atlantic loop to be in the billions of cubic meters per hour.

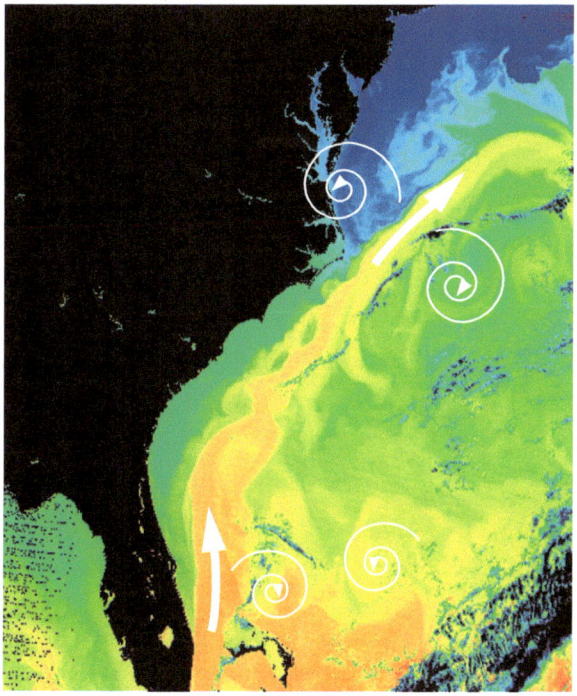

Figure 5.8 - False-color photograph of the Gulf Stream along the coast of North America, taken by NASA. Note the higher temperature (orange) to the south-east of Florida and the formation of vortices (yellow) in the somewhat cooler waters (green). Vortices are also visible in the much colder waters from Cape Hatteras in North Carolina to those of New York (blue). [© Donna THOMAS/ MODIS Ocean Group NASA/GSFC SST produced by R. EVANS *et al.*, U. Miami]

To complement figure 5.7, figure 5.9 illustrates the large diversity of well-known surface currents, which can be considered as detours caused by deflections against the continents. Some of them can be clearly identified, such as the HUMBOLDT Current, which is part of the South Pacific Gyre that is blocked by the west coast of South America and turned northward. Without describing all these currents, a quick overview of figure 5.9 reveals the influence of the coasts, which can block a circuit, reorient it, and generate secondary loops.

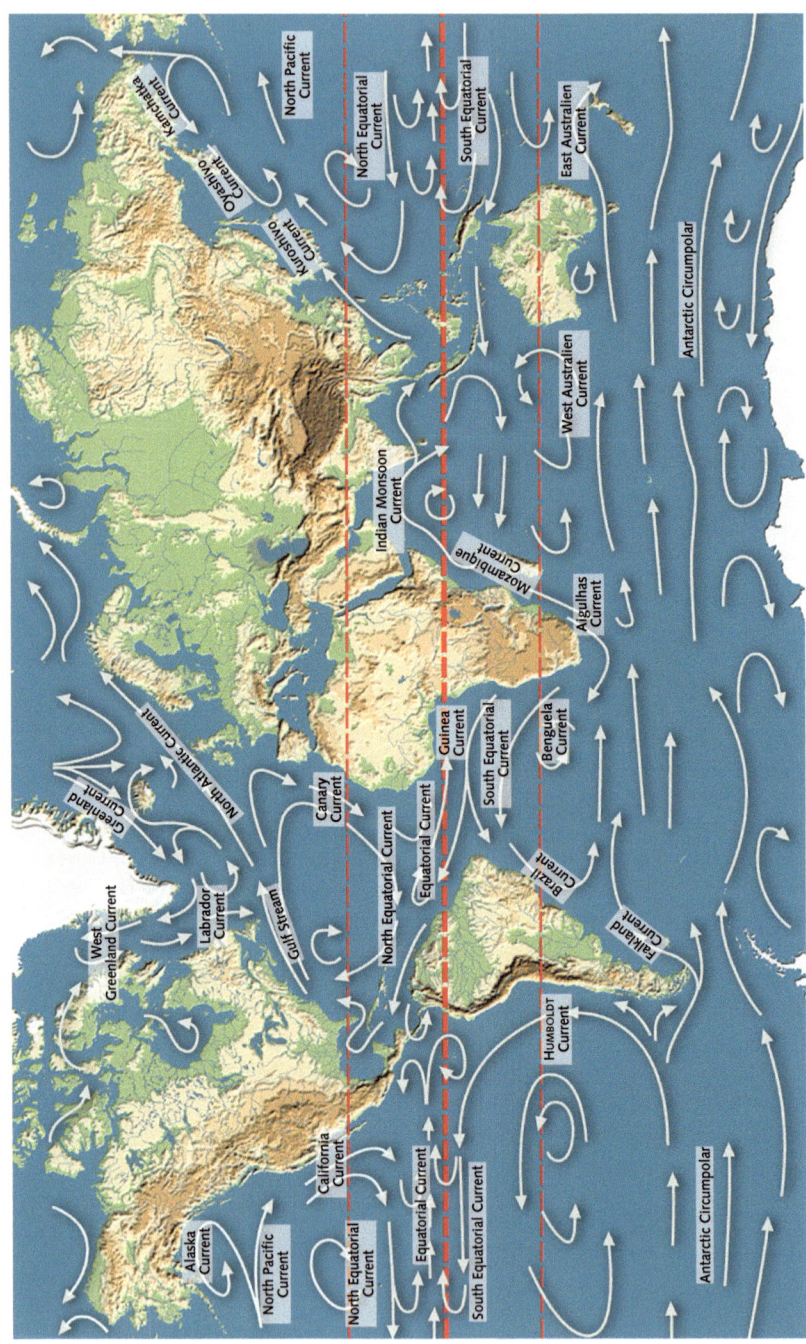

Figure 5.9 - Illustration of the diversity of catalogued oceanic currents that flow around the thermohaline circulation represented in figure 5.7. [© Yuvanoe / CEA]

Conclusion

The *vast reservoir* that is the ocean, to paraphrase Jules VERNE, represents much more than the aqueous immensity that we have described. The sea is also the cradle of life. We have only observed the main physicochemical properties and explained why the sea can be sailed with rather moderate risks by so many surface ships. River water and precipitations enter the sea charged with all sorts of suspensions whereas, via evaporation, only fresh water escapes. All the contaminants thus accumulate in the oceans, which have an extraordinary capacity to absorb them, using some of them to nourish the animal and plant species that live there, dissolving others into the sediments, which could constitute the crust of continents of a geological era still to come.

Even in the calmest conditions, the sea is never stationary. It is everywhere traversed by powerful oceanic currents, even if they are extremely slow, especially in the deepest portions of the loops of the global thermohaline circulation. We have seen that this main circuit determines all the others. Its slowness justifies its description in the first sections of this chapter as quasi-static. Nevertheless, as slow as this oceanic river may be, the flow rates transported are such that they exerts a significant influence on the atmospheric circulation and meteorology. The seas that lie beneath the surface of this gigantic circuit are far from being flat or round but have a structure that is due essentially to the local variations in gravity.

Chapter 6

The sea that we see dancing

Waves scree and dare, leaping from rock to rock.

(Paul VALÉRY, *The Graveyard by the Sea*)

© Springer International Publishing AG 2017
R. Moreau, *Air and Water*,
DOI 10.1007/978-3-319-65215-3_6

U nder the apparent tranquility of the sea is concealed an entire palette of oscillations and pulsations, some susceptible of becoming extremely violent. Certain ones obey a relatively clear mechanism that we shall analyze through our simplifying lens, preferring to guide the reader so that he can evaluate the potential consequences of these phenomena, rather than providing a detailed scientific explanation. We begin with observations of the tides, which are almost periodic, and whose main characteristics are relatively simple to describe, then follow up with the phenomena of *El Niño* and *La Niña*, which are responsible for climatic anomalies that have now been identified, but whose exact causes remain debated. Finally, we examine swells—oscillations that are rarely calm and that always carry a colossal energy. We shall also overview all sorts of waves in a quest to understand why some can become so menacing and destructive.

1. The tides

In twelve hours, the Moon orbits about half-way around Earth, but each day's trajectory differs slightly than the previous days'. Consider first when the Moon is at its zenith over the coastline of a given ocean. It attracts and lifts the water near this coast, leading to a lowering of the water level on the opposite shore by conservation of mass. Several hours later, the Moon appears over a different coastline and raises the water there, and the level lowers again along the first coastline. We can thus understand that an oscillation is established on the scale of the oceans (tens of thousands of kilometers) that displaces colossal masses of water and generates rather high speeds that are relatively easy to evaluate. Tides thus appears as the oceans' response to the periodic excitation by the attractive force exerted on their mass by the Moon, but also, as we shall see, by other celestial bodies, such as the Sun.

The amplitude of this phenomenon can be particularly high in the oceans because they are of sufficient size to have their own, intrinsic periods[1] that are rather close to those of the nearest celestial bodies. As holds for any liquid confined in a basin, the mass can oscillate naturally as almost one unit, alternately climbing on one side and falling on the other like a sort of liquid pendulum, independent of the exciting mechanism. The order of magnitude of the natural period T can be evaluated with the simple formula for a pendulum $T = 2\pi\sqrt{L/g}$, where L is the width of the

---

1   The intrinsic period of a mass of water in a basin of length l is its period of the global oscillation, which shifts the water almost as a single unit back and forth from one extremity to the other. This is exactly analogous to the period of a pendulum of length l, which is $2\pi\sqrt{l/g}$. With $l \approx 10\,000$ km (as for an ocean) and $g \approx 10$ m s^{-2}, we obtain a period on the order of 2 h. With $l \approx 1$ m (as for bathtub), this period falls to 2 s. These numeric examples show clearly that immensely large basins are needed to obtain periods of oscillation that are comparable with the lunar period.

basin or of the ocean and g is the acceleration on Earth due to gravity. In a garden bucket, where L is of the order of a meter, this formula leads to a period of the order of seconds. However, in the ocean, where L is of the order of tens of thousands of kilometers, this formula leads to a period of the order of several hours, which is similar to the intervals between the transits of the Moon.

If the period between maximum height of the natural oscillation at one extremity of the basin coincides with the period with which the Moon appears over this same extremity, then the attraction of the Moon is exerted at the best possible moment for this phenomenon to reach its maximum amplitude. In this case, we say that the Moon's transit is in phase with the natural oscillation. All ex-children who can remember feeling just the right moment on a swing to apply the kick to maximize the amplitude of the oscillations should have no difficulty in understanding this optimization mechanism. In effect, tides are a manifestation of a gigantic oscillation of the mass of water contained in an ocean, which tends to follow the celestial bodies that attract them alternately from one side to the other. We say that the system is in resonance when the period of excitation, in this case of the attraction of the water by the Moon (or also the effort of the child on a swing), coincides with the period of the ocean (or the swing). In each case, the resonance leads to the maximum amplitude of oscillation.

When the excitation period is close to the oscillation period but does not coincide, as is the case for the largest oceans, which are influenced by the Moon and Sun, the amplitude of the phenomenon is still high. However, the maximum decreases rapidly as the difference between the periods increases. In fact, on the scale of several hundreds of kilometers, such as for the Mediterranean Sea or the Great Lakes of North America, the Moon and Sun are seen at the same time almost everywhere, so that, if their attractive force is present, it acts as one on the entire liquid mass in these reservoirs and is counterbalanced by its weight. The attraction by these bodies on these liquid masses of moderate dimensions thus cannot lead to shifts as large as for the oceans, although the phenomenon is detectable. On even smaller scales, such as for the lakes that dot our countrysides, fortunately no tides are visible.

Anyone who strolls along the French Atlantic or Normandy coasts has noticed the considerable variations in tidal range[2] from one day to the next, from one season to the next, or even from one port to the next. The phenomenon is apparently not as simple as the preceding lines would have us believe. In the first place, all celestial bodies attract the ocean waters with a force that is proportional to its mass and inversely proportional to its distance squared, as stipulated by NEWTON's law of universal gravitation (see chap. 2). It is true that the attraction of the Moon, which is near but not very massive, is the most important contribution. The contribution

---

2   The tidal range is the difference between the sea level measured at high tide and at low tide. It is thus twice the amplitude.

of the Sun, which is much more massive but also much farther away, is about half that of the Moon, which implies that it must be considered by whomever wishes to correctly predict the time and magnitude of the tides. In comparison, the contribution from the other planets of the solar system and of all the other celestial bodies is much smaller, so typically they are neglected.

In the second place, all the large movements of the open seas and oceans are coupled because of the continuity of this immense marine system, so that no ocean behaves exactly as a closed basin. Thus, on the scale of Europe, the Atlantic tide can easily rush into the English Channel, which is wide open between the French and English coasts, but also into the North Sea. This double withdrawal constitutes a significant perturbation that prevents the Atlantic tide from faithfully following the oscillatory properties of water in a closed basin. In these two narrow and shallow seas, friction significantly attenuates the tides, weakening them to the point that, where they meet on the western coast of Denmark, the resulting tidal range is much smaller than that of the Atlantic Ocean along the French or English coasts.

Now that we are well familiar with the geometry of the coasts and the topology of the seafloors, and with the trajectories of the celestial bodies nearest Earth, the relevant authorities are able to predict the tides with very good precision: the equations and the boundary conditions are well established, the numerical algorithms to solve them are finely tuned, and computers are sufficiently powerful to do the calculations. Thus, in each port of Brittany or Normandy, the sea level is known with precision several months in advance. The relevant services thus have ample time to adapt the schedule for the various comings and goings of the ships that serve the neighboring islands to ensure sufficient water depth in each port for maneuvering the boats. These schedules are known and published several months in advance.

Let us attempt to discern the origin of the variability of the tidal ranges due only to the fact that the Moon and the Sun follow their own orbit,[3] with their own period, without going into the details of this exercise, which would quickly become overly complex. We must distinguish between three distinct phenomena. First of all, the period of the Moon, which is the time required for it to orbit once around Earth, is 24 h, 50 min, and 28 s. Although close, it does not exactly match the length of an Earth day (24 h), which is the time it takes for Earth to rotate once about its axis. Second, everyone knows that, during a new Moon, our satellite passes between the Sun and Earth, which explains why it is not visible from Earth. To be precise, the period between the simultaneous passage of the Sun and the Moon above the same meridian is 29.53 days. This is the synodic month, or the duration of a lunation, which does not coincide with the average calendar month, although it is close. Third, the Sun passes about twice a year through the plane of the lunar orbit, which is the most propitious period for eclipses. Here again, we can be more

---

3   In reality, the Moon orbits Earth, but it is Earth that orbits the Sun and not the opposite: GALILEO was right: *E pur si muove!*

precise: the time between two successive transits of the Sun through the plane of this orbit is about 173.31 days, which differs little from half of a tropical year (182.62 days; half of a tropical year separates equinoxes, which is when the Sun passes through the equatorial plane of Earth). The proximity of these times can be illustrated by the consequences: the eclipse of the Moon on January 9, 2001 was followed by an eclipse of the Sun on June 21, 2001.

This description of the phenomena imposed by the systematic differences between the period of the Sun and that of the Moon, even if supported by several pertinent figures, is too incomplete for us to understand in detail the interactions between Earth, the Sun, and the Moon. However, it suffices to justify the conjunction between the phenomena that can reinforce or oppose each other, or can shift only partially, thereby generating the strong variations in tidal ranges.

Let us complete this description by situating the best-characterized extremal conditions, such as those illustrated in figure 6.1. The attractions of the Sun and the Moon add together when these two bodies are aligned with Earth (position 1); the overall effect is thus maximized. On the contrary, the attractions cannot combine when the Earth-Moon axis is perpendicular to the Earth-Sun axis (position 2). The alignment of the three bodies Earth, Moon, Sun is optimal during the solstices, when the Sun is above one of the tropics. In this situation, the solar attraction is maximized on one of the tropics and minimized on the other, so that the period of the solar tide is about 24 hours (the time for a single revolution of Earth on its axis). To illustrate this situation, recall that, in the Northern Hemisphere, strong tides with a period of 24 hours occurred on January 5, 2001, which is close to January 9, 2001, the date of the lunar eclipse and the winter solstice. The same occurred on June 28, 2001, which is close to June 21, 2001, the date of the eclipse of the Sun and of the summer solstice.

The equinox tides, reputed for their extreme tidal range, have a different explanation. At this time of year, when a point on the equator sees the Sun at the zenith, the point also on the equator but on the opposite side of Earth is the farthest from the Sun; thus, the least attracted. This second point will, in turn, see the Sun at the zenith 12 h later. The solar tide is thus maximized twice during one revolution of Earth about its axis and thus has a period of about 12 h. The high equinox tides, well known to the inhabitants of ports such as St. Malo, occur when the maximum of the solar tide coincides with the maximum of the lunar tide. Another case altogether occurs during the solstices, because the Sun is above one of the tropics, where the tide is maximized, whereas the tide is minimized under the other tropic. The solar solstice tide is thus maximized one single time over all of Earth, so its period is 24 hours. Outside of the equinoxes and the solstices, the two periods of 12 and 24 hours both make themselves felt and lead to two components of the overall tide, which leads us to distinguish in each port the diurnal tide and the semidiurnal tide.

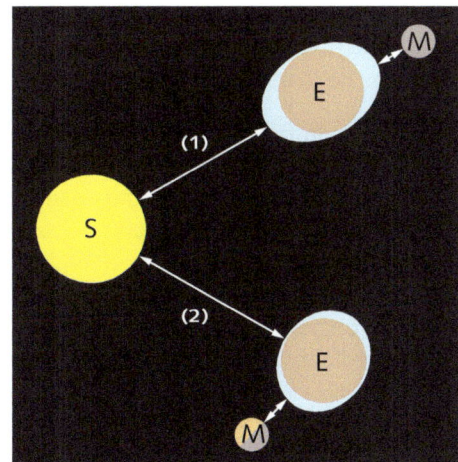

Figure 6.1 - Tidal mechanism (tides shown in light blue): maximum when the Sun (S), Earth (E), and the Moon (M) are aligned (position 1) and minimum when the Earth-Moon axis is orthogonal to the Sun-Earth axis (position 2).

In the preceding paragraphs, in trying to reduce the phenomenon of the tides to its dominant characteristics, several complementary effects were omitted, some of which merit being mentioned. So start with, considering the relative orbits of the Sun, Earth, and the Moon, another very long cycle of about 18.6 years is superposed over those evoked above as well as over the annual cycle. Thus, during 9.3 years, one of the two hemispheres is closer to the Sun than the other, so the sea level rises several tens of centimeters in the given hemisphere and decreases by the same amount in the other. This was the case for the Northern Hemisphere from 2005 to 2015. The inverse will be true for 2015 to 2025. Although this perturbation poses a problem for the specialists that monitor the sea level, primarily to estimate the fraction of its rise that is due to global warming[4] (notably via the data provided by the Topex-Poseidon, Jason, and GOCE satellites; see chap. 5), they are perfectly able to adjust for it.

Finally, at the scale of a sea such as the English Channel, which is too small to have significant tides but finds itself periodically invaded by the tides from the Atlantic Ocean, the apparent phenomenon is not a unified oscillation of its entire volume of water but alternating currents that go from the Atlantic to the North Sea and vice versa, with a period imposed by the Atlantic tides. When present, the current that travels from the Atlantic to the North Sea is deviated by the CORIOLIS force to the right toward the French coasts, which block it. To push the current back out to sea while maintaining its eastward momentum, the coasts are subject to an additional uplift of the water. Reciprocally, the return current toward the west, which is also deviated to

---

4   Any warming of the seas causes their water level to rise because of two different mechanisms: The first, which explains about 65% of the phenomenon, is the dilation of water as its temperature rises. The second, which accounts for complementary influences, is the partial melting of the glacial ice caps on the continents, essentially in Greenland and Antarctica. Because of ARCHIMEDES' theorem, the melting of the sea ice cannot lead to a rise in sea level because the floating ice displaces exactly the same amount of water that it produces by melting.

its right, sucks water away from the French coasts, locally lowering the water level. This explains why the tidal range is much greater on the French coasts (up to ten meters, depending on the season) than on the English coasts (only a few meters).

Moreover, in the Bay of Mont St. Michel, another rather particular effect is super-posed over the CORIOLIS force and amplifies even more the local tides. This is the reflection of the tide from the Atlantic off of the western coast of the Cotentin peninsula. The topography of the coasts is such that the reflected wavefront arrives in the area of Saint Malo almost in phase with the Atlantic tide, itself already ampli-fied by the CORIOLIS force. Each year in September large crowds[5] are attracted to view the spectacle of the high equinox tide, which is added to the two effects discussed above.

2. The *El Niño* phenomenon

In spite of its considerable inertia that renders it practically unaffected by the rapid oscillations of the atmospheric airflows, the large-scale oceanic current is not exactly stationary. The example of quasi-periodic oceanic current (albeit with a strong random component) that draws the most attention is certainly the *El Niño* phenomenon. This current, which is sometimes referred to as *ENSO* (*El Niño* Southern Oscillation), is responsible for significant meteorological phenomena. Its origin is found in the atmospheric oscillations in the WALKER cell, which is described in chapter 2. When this cell is well formed between the tropics and is relatively sta-ble (fig. 6.2, top panel), the westward wind over the surface of the Pacific is much stronger than the westward equatorial wind. The surface water is thus very effi-ciently swept along by friction and is increasingly heated by the Sun. Upon meeting the coasts of Australia and New Zealand, these waters are diverted northward and southward, thus supplying the local marine currents. The water that reaches the west also warms the local depths, thereby lowering the thermocline. In contrast, the lack of water in the east causes a local lifting of the thermocline and an increase in phytoplankton, which is accompanied by numerous animal species. Thus, in the stable regime and on the scale of the Southern Pacific Ocean, the thermocline is significantly higher in the east (South American side), where fish are abundant, than in the west (Australian side), where the temperature is markedly higher.

Associated to these temperature differences is the convective circulation of the WALKER Cell in the troposphere. Because we are in tropical latitudes, the warm air charged with moisture is lifted above the eastern Australian coasts, where it cools and dries in the high troposphere, then redescends above the South American coasts. A contribution of cold, dry air from the high layers of the troposphere thus

---

5   You may be familiar with the old adage according to which the tide in the Bay of Mont St. Michel advances at the speed of a galloping horse.

establishes itself over the western tropical regions of South America (Columbia, Equator, and Peru). Moreover, farther south, the western coast is cooled by the HUMBOLDT Current that carries cold Antarctic water northward along the Chilean coasts toward the tropical latitudes. This global system, combined with the high Andean plateaus, explains the dry, cool climate that is quite typical of this part of South America.

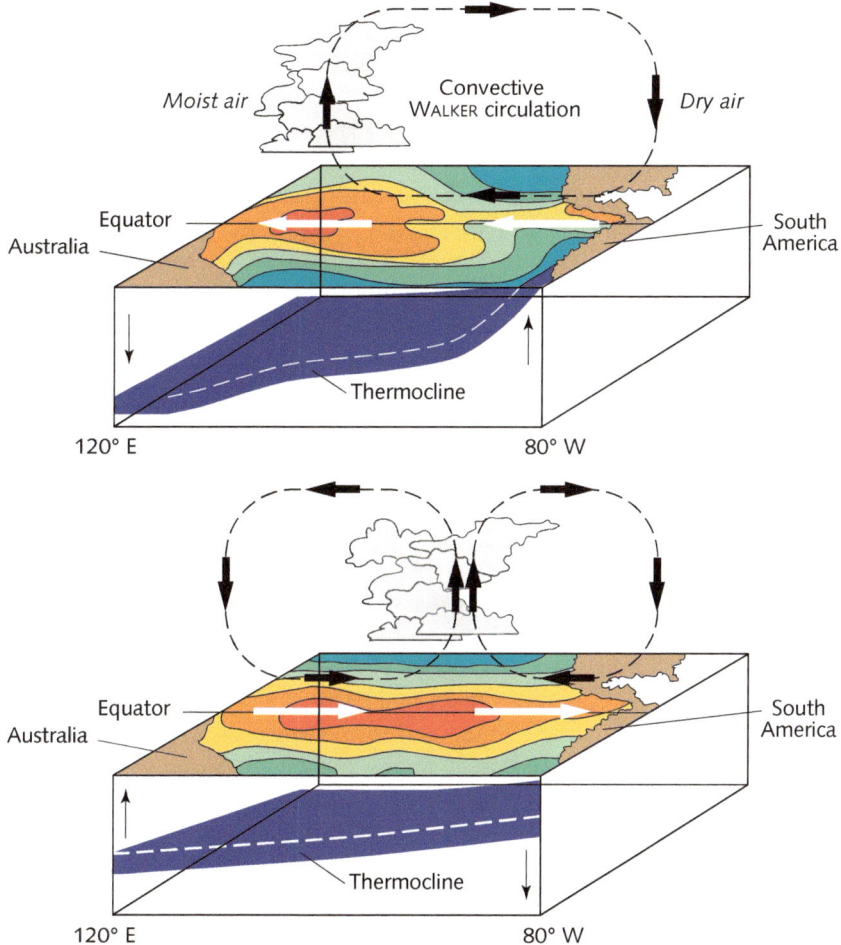

Figure 6.2 - Illustration of the interaction between the WALKER cell and the *El Niño* phenomenon. In the upper panel, the WALKER cell is active and accentuates the westward airflow, pushing the warm waters toward the western coasts of the Pacific. In the lower panel, the WALKER cell is broken down and divided, so the warm waters accumulated in the western Pacific flow back toward Central America to the east. The arrows ↑ and ↓ show the shift in the thermocline. [from the TAO project Office (director Dr. Michael J. MCPHADEN)/PMEL/NOAA]

Because of complex mechanisms[6] that still animate discussions between specialists, the WALKER cell loses its intensity at almost-periodic intervals, to the point that
it disappears and reconfigures itself differently (fig. 6.2, lower panel). Thus, the
westward surface current in the Pacific Ocean cannot be maintained, and the previous hydrostatic equilibrium within this ocean is upset. The warm waters accumulated to the west flow back toward the east, where they modify rather strongly the
water temperature and perturb the marine currents. The warm currents directed
toward the coasts of Peru and Equator are called *El Niño*. The thermocline rises
in the west and drops in the east. In the high atmosphere, the WALKER cell cannot
maintain its form and is replaced by two smaller cells (fig. 6.2, lower panel). This
new organization strongly modifies the normal regime that is characterized by a
high-altitude wind out of the west, which is constant and dry when it arrives in
South America. The double cell represented in the lower panel of figure 6.2 is,
on the contrary, capable of dismantling the high-pressure zone over the Easter
Islands, which is fairly typical in the middle of the Pacific Ocean, of stocking up
again with moisture, and of bringing strong precipitation to South America, with
violent storms and catastrophic flooding. The duration of *El Niño* is of the order of
18 months, but with rather strong variations that are difficult to predict.

The same phenomenon but of opposite sign, which corresponds to an amplification of the WALKER cell, also exists, even if it is generally less pronounced than its
breakdown. It is called *La Niña*. Thus, the westward transport of warm waters is
enhanced. To the east, the thermocline rises even higher than during the stable
regime, the waters are even colder, and the flow of dry, cold air arriving on the
South American coasts increases. The marine current toward Australia is amplified.
On the western coasts of South America, this phase of the oscillation is accompanied by a very large bloom of phytoplankton, a particularly good fishing season,
and an accentuated dryness. The *ENSO* is now considered to be an extremely
important phenomenon because of its meteorological and climatic repercussions,
which are due to its large amplitude and to the immense length scales involved.
Its worldwide influence is accepted, to the point that it can also influence the
European meteorology, over 20 000 km away. This phenomenon thus continues to
be investigated, based notably on the data supplied by the satellite Jason.

This overview of one of the best organized oceanic oscillations would not be complete without two remarks on the immense range of timescales over which they
occur. The shortest timescale is that of the tides mentioned in the previous section:
24 h. The longest timescale is the round-trip time of the thermohaline circulation:
1600 years, or about 10^7 times the timescale for tides. And, of course, the secondary currents that branch off of this circulation in the diverse oceans, as well as the
decelerating vortices evoked, bring into play all the intermediate scales. If nothing

---

6   One of the better-reasoned explanations is founded on the fluctuations of the HUMBOLDT Current,
     which carries cold water from the Southern Ocean. But it still appears difficult to separate cause
     and effect in the fluctuations of the coupling between this current and the WALKER cell.

else, recall only that the ocean is an extremely massive and very slow system with vary diverse periods and timescales.

3. Swells and waves

3.1. How and why do waves propagate?

Independent of the tides, which constitute the sea's response to the attraction of celestial bodies, and the large oscillations coupled between the ocean and the atmosphere, such as *El Niño* and *La Niña*, the surface of the sea is incessantly animated by the rhythm of the waves of diverse forms and wavelengths. These waves perturb the local height of the free surface and propagate with a certain celerity. The main driving force is the wind, which, somewhere, through friction, endlessly transfers a part of its energy and momentum to the seawater. This action is localized near the surface, extending down the thickness of a liquid layer, which is of the order of the wavelength of the waves. Thus, short waves act only on the water within a meter of the surface, whereas long waves accompany an oscillatory movement that reaches down to a depth of the order of tens of meters.

The driving wind is not necessarily the one blowing at the point of observation of the waves. In fact, the wind blows almost everywhere over the surface of the seas, exciting oscillations of the free surface and thus ceding some energy to the waves. The dissipation of this energy by viscosity, even when efficiently relayed by turbulence, is so slow that the waves can be transported over the entire breadth of the vastest oceans without being strongly attenuated. In addition, over such long travels, waves have the opportunity to encounter one or even several low-pressure zones, where the winds are capable of resupplying them with energy. Overall, the energy of the air-water partnership comes from sunshine, as we have already touched on. The movement of these fluid media is necessarily such that their energy, relayed by turbulence, can ultimately be dissipated by viscosity into heat, but generally this happens at very large distances from their point of origin.

Although the tides constitute movements of immense masses of water during periods close to 12 or 24 h, waves, on the contrary, must be seen as much more local oscillations because their wavelength is at most of the order of several tens of meters. They are also much more rapid because their period is of the order of seconds. They are gravity waves within which the energy of fluid particles constantly convert from the potential form within the field of Earth's gravity (close to the crest) to the kinetic form (between the crests). They displace water over a closed trajectory on the scale of their wavelength and during their period. Any water particle that we care to mentally isolate only makes round trips around its closed trajectory, without ever traveling large distances. In contrast, on the scale of a sea or an ocean, waves transport both their form and their energy by using water as the supporting material, indispensable to their transit, but without causing a net flow: anyone can watch the endless march of waves toward the shore without ever being flooded.

If we attempt to photograph at given intervals groups of waves that are separated by an integer number of periods, and then superpose the photographs, we find that their form is never reproducible but constantly evolves. When the sea is relatively calm, we can nonetheless notice a mean morphology closely resembling a sinusoid. In contrast, under a particularly intense low-pressure zone, which contributes to generating waves by providing them with energy, the wave morphology is very irregular and chaotic, in general marked by breakings. However, with increasing distance from such a zone, the waves progressively stabilize and effectively evolve toward an ever-more-regular morphology. What are commonly called *swells* is a system of waves that have become sufficiently uniform so as to appear almost regular. These can be observed both far from the point where the wind creates them and far from the coasts (fig. 6.3).

Just as for light and sound waves, swells can be analyzed with spectral methods, which reveal that, despite their apparent regularity, they never form pure oscillations that can be characterized by a single period. Under the simplest conditions (far from where the swells are produced and in a very deep part of the sea), these oscillations are quasiperiodic and are composed of modes whose wavelengths and periods that are fairly close to an average value. If, under such conditions, we measure and analyze them in detail, we can verify that the spectral distribution is relatively narrow around a principle mode and its harmonics.[7] Recalling some vocabulary already seen in chapter 1, we can say that the spectrum of the swells stabilizes very close to the spectrum of a single ray.

To simplify the description of a relatively uniform swell, let us assume that they can be reduced to a single mode; in other words, a single, sinusoidal oscillation with wavelength λ, period T, and celerity $c = \lambda/T$. Figure 6.4 shows the paths followed by fluid particles during the passage of such a wave, which leaves the particles near their starting point. These paths are nearly circular for waves whose amplitudes are roughly a half wavelength and whose wavelength is very much less than the water depth, so that the effect of the seafloor can be neglected, as if it were infinitely far away. Conversely, for wavelengths comparable to the water depth, the seafloor exerts its influence, and the paths become more and more elongated into a flattened elliptical form. In both cases, the fluid particles undergo zero net displacement because the paths are closed.

---

7   A harmonic is a component of a periodic oscillation; its frequency is an integer multiple of the frequency of the principle mode. Thus, sound with frequency f is generally accompanied by sounds of frequency $2f$, $3f$, etc. Sounds of frequency $f/2$, $f/4$, etc., also exist and are called *subharmonics*.

(a)

(b)

Figure 6.3 - (a) Small swells whose amplitude increase as they arrive in a shallow zone, where they finally break. Even in the open water, where swells are relatively uniform, their moderate but systematic separation is noticeable from one point to the next [© SIDS1/Flickr]. **(b)** Near a low-pressure zone, strong waves that crest several meters high and with frequent whitecaps forming in the open water [© Olivier DUGORNAY/ Ifremer]. In both cases, the breaking of the wave is marked by a white, short-lived foam.

In a first approximation, the celerity[8] of these waves, which is the speed at which their shape and energy propagates, adjusts so that the variations in kinetic energy

---

8   Assuming that the maximum speed of the water particles is approximately the same as the celer-
     ity c, then we are led by the equality between the kinetic energy of a unit volume of water

compensate those in potential energy, as happens for a simple pendulum. But this simplified explanation masks other phenomena, such as the dispersion of these waves that do not all travel at the same speed, or the interactions between modes that redistribute energy throughout the entire spectrum. Even if only approximate, this simplification has the advantage of leading to the order of magnitude for the celerity, which in deep water is between 5 m s^{-1} for short waves and 50 m s^{-1} for long waves. These quantities show that long waves travel faster than short waves and can therefore catch up with them. This constitutes one of the characteristic properties of dispersive waves. We can see that, if the long waves are capable of catching up with short waves, the former can also capture the energy of the latter, thereby growing even larger. Thus, under a vast high-pressure zone, far from low-pressure zones where winds can pump energy into these waves, swells can appear as a quasiperiodic and quasimonochromatic phenomenon.

If the formation of swells is essentially due to a transfer of momentum and energy from the wind to the sea via friction, other mechanisms, such as the strong pressure variations over relatively short distances such as for cyclones, complicate the dynamics of this interaction between air and water. Nevertheless, once again, it is possible to extract several simplified notions. All else being equal, the wave height increases the longer the wind blows, because this gives the wind ever more time to transfer its energy to the water. In the same vein, the waves become higher with increasing distance over which the wind blows,[9] because their amplitude grows progressively as they cross the low-pressure zone. To obtain several characteristic orders of magnitude, recall that a force 7 wind[10] blowing for 10 h over a 100 km zone of water can generate swells with 3 m troughs, which are quite typical of the Atlantic Ocean. In the Pacific Ocean, the typical dimensions of the water surface and the low-pressure zones can be much larger, which means that we can see swells with troughs on the order of tens of meters, which are sought out in particular by the surfers on the beaches of Hawaii.

$(\rho c^2/2)$ and its potential energy (which can be estimated as $\rho g \lambda/4\pi$) to the formula for celerity: $c = \sqrt{g\lambda/2\pi}$. Here, λ is called the *wavelength*, and $\lambda/4\pi$ is a good estimate of the height through which the fluid particles move. This result is applicable to waves in deep water; in other words, where the depth h is significantly greater than the wavelength λ. However, in the opposite case, such as in lagoons and on beaches where $\lambda \ll h$, the formula for celerity becomes $c = \sqrt{gh}$. In fact, these two expressions are asymptotes of the more general formula given by the theory of sinusoidal waves:

$$c = \sqrt{\frac{g\lambda}{2\pi} \tanh \frac{2\pi h}{\lambda}} \; .$$

9   Sailors use the word *fetch* to designate the distance on the ocean over which a wind blows with constant velocity.

10 To characterize wind speed, sailors use the BEAUFORT scale, according to which a force 7 wind blows about 30 km h^{-1} (~ 20 mph) and forms breaking waves. The maximum is a force 12 wind, which blows at over 120 km h^{-1} (~ 75 mph) and is associated with a genuine storm.

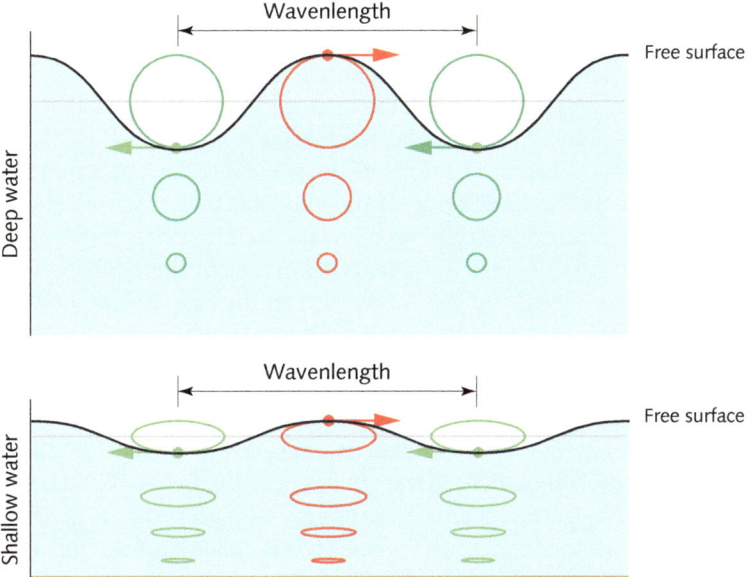

Figure 6.4 - Closed paths (green and red arrows) showing the alternating circulation of water particles due to the passage of a supposedly sinusoidal wave. Top panel shows waves with short wavelength compared to water depth; paths are quasi-circular. Bottom panel shows waves with long wavelengths; paths are flattened.

3.2. The surprising variety of waves

Figure 6.3a shows a situation that anyone may see from a beach. It shows a progressive growth in wave height as the water depth diminishes, as well as an uplifting of the crests of the waves, which leads to their breaking. Consider first the growth of the wave amplitude. This is due to the fact that, once a wave encounters the seafloor, the local instantaneous flow rate that circulates under the wave has no choice but to push the water up into the air, following which the water must return to the free surface. This rising of the crests above the free surface thus results from the presence of the progressively growing obstacle constituted by the inclined shore on which the overpressure acts to provoke the local lifting of the layer of water. This can only happen when the water is too shallow for the paths shown in the upper part of figure 6.4. The threshold depth at which the seafloor influence begins to be felt is of the same order of magnitude as the wavelength of the waves.

Although the proximity to the seafloor explains the wave amplitude, it does not explain why the wave crest lifts and the face steepens, which is visible in figure 6.3. This lifting and steepening is a classic property of the KELVIN–HELMHOLTZ instability, according to which the vorticity concentrates near the inflection point situated on the leading face of the deformed interface. In the appendix, this property is explained for the special case of a shear interface in a single-fluid medium, but it is also true when the fluids on each side of the interface differ, as is the case for surface waves on the sea. To keep things brief, let us say that this property results from the fact that the crests, where the wind speed is maximal, move faster than the troughs, where the wind is minimal, so that the crests catch up with the troughs. In the vicinity of the troughs, however, the wind speed is minimal and its pull on the water is reduced. The slope of the wave face thus becomes steeper and steeper, whereas the backside slope of the wave becomes less and less steep. Put another way, the horizontal extent of the wave face becomes shorter to the point of becoming vertical, whereas the horizontal extent of the backside of the wave approaches the wavelength of the wave. We can thus understand how waves distort in shape as they approach the beach, as illustrated in figure 6.3b. We will again encounter this phenomenon a little later when we discuss tsunamis (see fig. 6.8).

The final stage of these two effects (i.e., the increasing amplitude and the uplifting of the wave face) is the break. It happens when the local slope of the wave face approaches infinity. Although moderate in figure 6.3b, the break can become spectacular, forming the classic tubes so sought after by surfers (fig. 6.5). They can also be dangerous, both for people and for small boats, when dealing with a large swell. In contrast, high-tonnage ships, such as cargo ships, supertankers, or even aircraft carriers, which can be longer than 200 m, barely notice swells whose wavelengths reach tens of meters and whose amplitude remains on the order of several meters.

Figure 6.5 - Example of a large-amplitude wave that can form a tube when breaking. Such waves are actively sought out by surfers. Note also the formation of the large amount of white air-water emulsion, as well as the spray. [© Misty/Flickr]

When a swell arrives near the coast, several phenomena become noticeable that we cannot describe in detail here, but that always produce an impressive spectacle. Let us begin by recalling that the reflection of waves against a vertical embankment can lead to a system of stationary waves due to the superposition of the incident and reflected waves. In these conditions, we can distinguish a succession of nodes and antinodes; in the antinodes the bumps and troughs follow each other periodically with the whole remaining practically immobile, whereas the surface does not move at all on the line of nodes. If the waves are incident at an oblique angle with respect to the embankment, we can observe a sort of checkerboard pattern on the surface, analogous to dappled clouds. This form is typical of two wave trains that cross each other. This checkerboard pattern can also be seen along the banks of canals traveled by barges (fig. 6.6).

This figure also shows a mechanism that produces waves that are completely independent of any eventual wind: the bow of the barge separates and lifts the water so that the boat can take its place. The figure shows well the separation between the agitated surface behind the bow wave on the one hand and the calm water of the canal on the other hand that, like a mirror, produces a clear reflection of the trees growing on the banks of the canal. The analogy between this bow wave and the shockwave that precedes an airplane in supersonic flight (see chap. 2 and fig. 2.8) merits a more detailed comment. At sea, as on lakes, all objects that move with respect to the water in fact generate waves whose regular structure we have all noticed. This water, which is raised up to a height that is slightly less than the draft of the hull, accelerates downward, thereby transforming its potential energy into kinetic energy. Once again, a sort of liquid pendulum is excited, whose oscillations are visible on each side of the vessel. When a swan swims peacefully across a still lake, the physical phenomenon is identical, even if the scene may seem poetic. In both cases, the waves formed on each side spread out from the point at which they were created and produce a clear line separating calm from wavy water.

Figure 6.6 - The yellow lines show the angle that separates the region containing the waves generated by the barge from the upstream region where the free surface, which is as smooth as a mirror, reflects clearly the tree branches and the bow of the barge. The angle α satisfies the formula $\tan \alpha = c/V$, where c is the celerity and V is the speed of the barge. Note also that, behind the barge, downstream of the blue line, the reflection of the waves from the embankment of the canal creates a checkerboard pattern that is rapidly perturbed by turbulence. [© Alan ABELLEYT]

Thus, an acute angle exists that separates the area where the free surface is agitated from the area which remains wave free. What is this angle? The waves that spread out the farthest are those that propagate with celerity c perpendicular to the direction of travel. If the vessel advances with speed V, such as the barge in figure 6.6, the angle α between its direction of travel and the edge of the area affected by the waves (i.e., the wake line) satisfies the formula $\tan\alpha = c/V$. This is always the case, in fact, when the speed of the vessel is greater than the celerity of the waves: the ratio c/V is less than unity, so the preceding equation has a single solution. This gives the angle α, which is remarkably similar to the half angle of the MACH cone (chap. 4, fig. 4.8).

What happens, however, when the speed V is less than the celerity c? The waves then propagate faster than the vessel in all directions, including the forward direction, so they invade the entire free surface of the basin. This is the case, for example, of waves generated by a stone thrown into calm water: the waves are generated by the uplift of the mass of water that is replaced by the stone and they propagate radially outward with celerity c, forming an essentially perfect circle. This also occurs around slow swimmers that tire themselves by struggling against the agitated pool water, not to mention the waves produced by other swimmers, and that risk swallowing water if they open their mouth the moment a small breaking wave arrives. In comparison, fast swimmers, who can swim faster than the celerity of gravity waves (i.e., less than 1 m s^{-1}) in a pool that is much deeper than the wavelength of the waves, are constantly protected from these perturbations because they travel faster than the waves.[11] A sort of small and very stable depression forms in the water around their head, or more precisely between their head and the hydraulic jump,[12] where the gravity waves cannot penetrate and which facilitates their breathing. This depression is the analog of the MACH cone, and the jump is the analog of the shockwave. Is nature rewarding the most deserving? Let us say, rather, that fluid mechanics offers rewards to swimmers that are capable of using its laws and properties.

When swells lose their regularity, other more precise expressions are often preferred, so rich is the vocabulary of the seas. Thus, experienced sailors know how to spot particular sequences of events, whether to use them or to protect themselves from them. For example, the fairly well-known phenomenon of the three sisters corresponds to a group of three waves, with the amplitude of the middle wave being larger than the preceding and subsequent waves, although without

---

11 The Brazilian swimmer César CIELO holds the world record for the 100 m freestyle with a time of 46.91 s, which means his speed was 2.13 m s^{-1}.

12 The hydraulic jump is the analog of the shockwave in aerodynamics: on the fluvial or subsonic side the perturbations propagate in all directions, whereas on the torrential or supersonic side they cannot propagate upstream because the source travels faster than they do. We again meet this discontinuity in the next chapter where we distinguish between fluvial flow and torrential flow in river systems.

being exceptionally large. We can also observe strange cases, such as rogue waves, whose amplitude can be two or three times larger than the preceding and subsequent waves, as is the case in figure 6.7. These can reach up to 30 m in height, and the overpressures that they generate can lead to extreme mechanical forces on the hulls of ships, sometime causing serious damage, or even shipwrecks.[13]

Other than being dangerous, large isolated waves have a curious property: they conserve their form and their amplitude over very long distances, without smaller waves being to perturb then in any significant way. They continue this way until they break, which is also shown in figure 6.7. The mechanisms that shape them are not simple, for they come out of nonlinear equations. In fact, this nonlinearity is precisely what allows the energy of an ensemble of small-wavelength, moderate-amplitude waves to become concentrated into a single wave, larger than the others and that continues to grow by capturing the energy of the slower waves.

(a)

(b)

Figure 6.7 - A rogue wave [© NOAA] **(a)** before breaking over the bow of the ship and **(b)** as it breaks over the bow.

---

13 Between 1973 and 1994, 22 cargo ships were sunk by rogue waves. In February of 1995, Captain WARWICK, commander of Queen Elisabeth II, commented upon seeing a 30-m-high rogue wave approaching: *I felt like I was heading straight for the cliffs of Dover.*

When the depth becomes less than the wavelength, certain large waves on the ocean surface acquire quite particular and remarkable structural properties, notably a robust stability that allows them to travel very long distances without losing energy. They belong to a class of much more general phenomena, which are called *solitary waves* or *solitons* in the physics texts that treat the theory of waves.[14] In fact, solitary waves occur in other classes of physical phenomena and we use them for various purposes, notably in the case of electromagnetic waves, whose applications in optics have proven to be of great significance. The modeling of solitary waves based on the equations of fluid mechanics[15] is now rather routine. Their observation in laboratories in large-scale wave channels has confirmed their stability and allowed their curious behavior to be studied in a reproducible way. For example, if two solitary waves are generated at opposite ends of a long canal and directed toward each other so that they are forced to cross, they reconstruct themselves almost identically after their short interaction with each other, except for a small loss of energy and amplitude. No sooner do they separate than each assumes again its initial form and continues its prior propagation.

Tsunamis are another category of extremely dangerous waves. They are created in the depths by a sudden shift of the seafloor over a rather large area, normally caused by an earthquake. This shift in the seafloor is analogous to the round trip of a giant piston, which shoves a large mass of water and endows it with enormous energy before returning to rest. The round trip of the piston triggers an oscillatory movement deep in the water, which is capable of propagating across even the vastest oceans. The initial speed and wavelength are imposed by the underwater piston. In the very deep ocean they diminish only very slowly because viscous friction on these immense scales is negligible (fig. 6.8). Often, tsunamis propagate near the bottom without being easily visible at the surface, before arriving in waters whose depth is of the same order as the tsunami wavelength, which can be as much as several hundred meters. Although they are initially invisible on the surface because the thickness of the perturbed layer at the sea bottom is of the same order as the wavelength, their influence attains the free surface when the depth diminishes sufficiently. They finally emerge in the form of several consecutive rogue waves as they approach the coasts. Just as for ordinary swells and for the same reasons, the amplitude of these waves can grow greater and greater and they can break on the coast, where they cause fearsome tidal waves.[16] The destructive power of 30-m-high wave is frightening.

14 The Scottish naval engineer John Scott Russell (1808–1882) appears to be the first to have seen and followed a solitary wave, which was triggered by the sudden stop of a barge in the canal connecting Edinburgh to Fort Clyde. He reports that the wave propagated without losing amplitude at the speed of his galloping horse.

15 When the depth is much less than the wavelength, the general fluid-mechanic equations can be simplified into the Korteweg-de Vries equation, so named in honor of the two Dutch scientists that first proposed it in 1895.

16 The word *tsunami* is more and more used instead of the word *tidal wave* to avoid confusion with the tides, which are also capable of causing flooding the land for several hours.

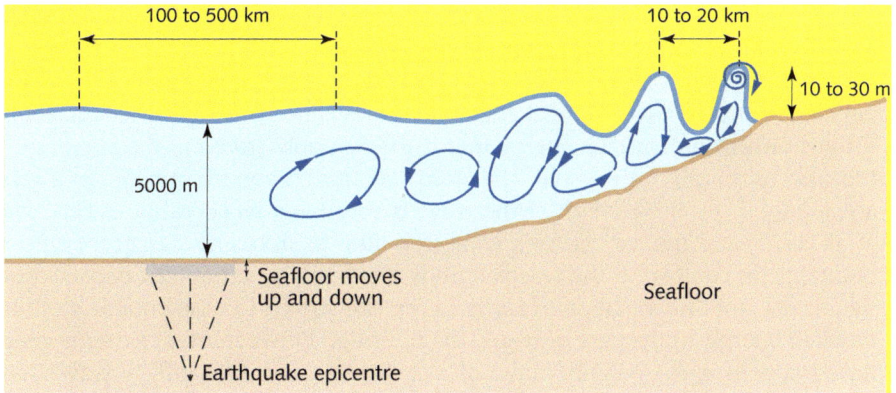

Figure 6.8 - Illustrative schematic and orders of magnitude typical of tsunamis caused in the depths by an earthquake. Note the growth in amplitude and the decrease in wavelength near the coasts, where the water is less deep.

The tsunami of December 26, 2004, caused by an earthquake whose epicenter was 250 km southwest of Sumatra and 30 km deep in the crust below the Indian Ocean, remains clear in our memories. Although the tidal wave was one of the most violent ever recorded, various fishermen report not having felt the tsunami as it propagated beneath their boats. The wavelength of these waves was around 30 km, meaning that they can be described by the gravity-wave theory in the shallow-water limit (for which the water depth must be significantly less than the wavelength). This explains why, in spite of its small amplitude in the open ocean, the tsunami can give birth to several waves some thirty meters high when it arrives near the coasts. In the open sea, the celerity of these waves is about 600 km h^{-1}. The tsunami of 2004 caused about 230 000 deaths.

On March 11, 2011, an earthquake of magnitude 8.9 on the RICHTER scale struck to the northeast of Japan, with the epicenter being about 130 km east of the port city of Sendai. It generated another catastrophic tsunami that led to some 20 000 deaths and over 100 000 refugees. It also seriously damaged the Fukushima nuclear power plant, triggering the release of radioactive pollution whose effects are, at the time of writing, impossible to quantify.

3.3. Ripples and convective instabilities under ice floe

Short-wavelength waves, on the order of 10 to 20 cm, are also detectable on the surface of water, even the calmest waters. They are often caused by a light, short-lived breeze and make up the chop that amateur sailors attentively watch for to detect any eventual changes in wind direction or intensity. In high seas, the chop forms waves with very short wavelengths (although still longer than about 10 cm) and superpose themselves onto swells without interacting with them in any significant way, since the order of magnitude of their wavelengths differ so much.

Another family of small waves exists with even smaller wavelength. These are *capillary waves* (often called *ripples*) and appear when the restoring force is no longer gravity, but surface tension (see chap. 3, insert 13.3). Instead of transforming kinetic energy into potential energy in a gravitational field, the kinetic energy is transformed into surface energy that is momentarily stored in the curvature of the crest or trough of the wave. The word *curvature* is important here: for a given amplitude, if the wavelength of the wave is very long, its curvature is very small at all points and the surface area varies too little to store any significant amount of energy. In contrast, if the wavelength is short, the curvature can become quite significant, and the variation in surface area can suffice to momentarily store the required energy. In the case of water, the transition between these extreme conditions happens in the neighborhood of a centimeter. Whereas gravity waves have wavelengths longer than tens of centimeters, capillary waves are recognizable by their short wavelength, which must be less than about a centimeter.[17] Thus, they can be seen in a bowl or even a glass. The celerity of capillary waves is of the order of cm s^{-1}, which is also much less than the celerity of gravity waves.

Independent of its aptitude for propagating short waves, capillarity is well known for the active role it plays in the formation of menisci near the line of contact between the free surface of a liquid and a solid wall, such as a glass. For water, the radius of curvature of such a meniscus is of the order of several millimeters. Capillarity is also what allows ink to ascend up the pores of a blotter, or for soil to soak up water.

The only significant part of the ocean that is free of waves is the ice pack, because a layer of relatively thick ice floats on the surface of the sea. The equilibrium of this layer, subject to its weight and the buoyancy force, implies that the ratio of the thickness above the water line to that below is about 0.087.[18] Can we say that no oscillation is possible underneath the frozen surface and that the water-ice interface is motionless? Two mechanisms act in opposition to this scenario. The first is the inevitable extension under the ice pack of the surface waves around the periphery. Periodic overpressures and underpressures slightly larger than those just external to the ice pack compensate for the fact that the ice does not immediately rise and fall with the free surface. Although friction along the water-ice interface is much greater than friction along a water-air interface, the viscosity of water is so small that waves manage to propagate and penetrate rather far under the ice before their energy is dissipated. Nevertheless, viscous friction is what explains the relative calm found in the cleared channels of a fragmented ice pack.

---

17 By equating the potential-energy density $g\lambda/2\pi$ with the surface energy $2\pi\sigma/\rho\lambda$ (where $\sigma = 0.07$ N m^{-1} is the surface tension of water), we obtain the threshold between gravity waves and capillary waves: $\lambda_c = 2\pi\sqrt{\sigma/\rho g}$. This value is about 0.016 m. The speed c of capillary waves can be found from $c = 1.5\sqrt{\sigma\lambda/2\pi\rho}$.

18 At the bottom of a layer of ice of thickness $t_a + t_b$ (where t_a is the thickness above water and t_b is the thickness below water), the pressure is the same as on the free surface of the water. This leads to the equality $\rho_i(t_a + t_b) = \rho_w t_b$, where $\rho_i \approx 0.92$ kg m^{-3} is the density of ice and $\rho_w \approx 1$ kg m^{-3} is that of water. From this we find the ratio $t_a/t_b \approx 0.087$.

The second mechanism for generating oscillations under the ice is the RAYLEIGH-BÉNARD instability (see appendix) within the layer between the thermocline and the frozen surface. The local density of sea water depends on both the temperature, being lighter at 0 °C and at 8 °C than at 3.98 °C, and the salinity (see chap. 5, fig. 5.2). The temperature difference has a stabilizing effect near the water-ice interface where the temperature hovers around 0 °C because light water finds itself in the higher regions. But this difference has a destabilizing effect in the deeper layers where the temperature can be greater than 4 °C, so that heavier water (near its maximum density) finds itself above lighter water. Moreover, because the ice constantly exchanges salt with the nearby water, both during its freezing as during its melting, the local variation in salinity also generates variations in density. During freezing, the water-ice interface ejects into the water a part of the salt that cannot be dissolved into the solid, where its solubility is less, thereby adding weight to the highest aqueous layers and destabilizing the system. During melting, newly formed water is lighter than deep water and tends to remain up high, stabilizing the system. The important thing to remember from this water-ice system is that both stabilizing and destabilizing mechanisms connected to the local temperature and salinity are present, but they do not act at the same depths. We can also add that they are not exerted over the same length scales, because the respective diffusivities of heat and of salinity differ by several orders of magnitude.

This complex system thus is neither stable everywhere nor unstable everywhere. Observations made underneath large icebergs that some plan to tow all the way to the Middle East to satisfy the demands for freshwater have shown that the interface between water and ice is far from a planar surface. The interface is undulated on the scale of the convective cells that develop underneath it. At some points the melting ice dominates and the surface is high, and at other points the freezing dominates and the interface is low. The distance between these high and low positions corresponds to the typical scale of the most unstable convective perturbations.

Conclusion

The seas seem to dance under the constant attraction of the Sun and the Moon, themselves subject to the rhythms of celestial bodies. Their changing reflection often suggests calmness and peace, but the force and intensity of the tides, inevitable and perfectly predictable, incite caution and reveal a colossal source of energy. This can be manifested by violent pulsations that result from changes in the marine environment. For the air, these changes may take the form of oscillations of the WALKER cell on the scale of the Pacific Ocean, which are capable of generating significant climatic perturbations, such as *El Niño*. On a more daily basis, and everywhere over the surface of the oceans, the meanderings of the atmospheric low-pressure zones supply energy to marine storms, which are sometimes accompanied by local perturbations that are particularly dangerous, such as rogue waves. Moreover, all underwater earthquakes arouse a response from the sea that is marked by events, some of which can reach catastrophic scales, such as the tsunamis in 2004 and 2011.

In all times, the sea has nourished its coastal inhabitants and has fascinated the innumerable wanderers along its shore. It has always attracted navigators and has been the theater of the Danish, Greek, and Roman conquests, before those of the great Spanish and Portuguese explorers. It has caused shipwrecks and tragic tsunamis. It has often been the dramatic theater of wars. But its beauty has always been sung by the poets, long before Charles TRENET, from whom the chapter title was borrowed. Somewhere, its power and the violence of its pulsations, often disguised under its apparent tranquility, contribute to this fascination.

Chapter 7

Rivers and streams

The water of the Nile loses its sweetness when it mixes with the sea.

(Egyptian proverb)

© Springer International Publishing AG 2017
R. Moreau, *Air and Water*,
DOI 10.1007/978-3-319-65215-3_7

T he continents are crisscrossed by numerous waterways of extremely diverse lengths and flow rates. They are supplied by precipitation and are subject to constant exchanges with both the water tables via percolation through the ground and with the atmosphere via evaporation and condensation. The springs, brooks, streams, canals, rivers, and reservoirs of all sizes participate together in the water cycle. Their highly branched distribution leads hydrologic engineers to divide the land into slopes separated by watershed lines. All of these waterways impact the geography and play a major role in the economy of the bordering communities. Some of the largest or most difficult to cross have been chosen to mark the border between neighboring states. Except for rare exceptions, such as the Great Salt Lake in Utah in the United States, they are characterized by the low salinity of the water that runs through them without dwelling. We will start our tip through this gigantic network with an overview of the large rivers, with special attention paid to identifying their importance. Next we will attempt to extract some properties of the flow within these waterways, highlighting the most remarkable phenomena, such as the hydraulic jumps, meanders, and waterfalls. We shall also touch on the economic and social role of this network.

1. The main properties of the great rivers

1.1. Length, depth, and speed distribution

The most-used criterion to classify the great rivers is their total length, from their source to their estuary; we have thus chosen this characteristic to compare in table 7.1 the 15 largest rivers on the planet. Their approximate average flow rate at the mouth is indicated in the right column and goes from 200 000 m^3 s^{-1} for the Amazon, which is 6150 km long, to 3100 m^3 s^{-1} for the Nile, which is 6670 km long. In comparison, the largest river in France is the Loire, which is 1012 km long and delivers about 600 m^3 s^{-1} to the Atlantic Ocean.

Each of these rivers possesses its own hydrological regime. As an example, consider the Yang-Tse-Kiang, which goes from high summer waters when it receives the abundant rains brought by the monsoons to a low water level in winter. For the largest rivers, however, local flood crests or low waters of their various tributaries take place in different seasons, so that, because of the complementary seasons, the appearance of these extreme phenomena can be relatively spread out near the estuary. Thus, the high waters due to snowmelt, which can be quite significant for rivers that drain large mountain ranges, are concentrated in springtime in the Northern Hemisphere. In contrast, the waters due to rain reach their maximum levels at the end of spring or in autumn. Nevertheless, a random element exists that can shift the timing of these maxima, and when high waters from different tributaries overlap, the downstream portions are subjected to atypical flooding. In contrast, each river in France (table 7.2) is much shorter and travels through a relatively small region, which means that the intermittency of their state can often affect their entire length.

Table 7.1 - The 15 largest rivers in the world, classified in decreasing order of length

River name	Countries traversed	Length [km]	Mean flow rate [$m^3 s^{-1}$]
Nile	Ethiopia, Sudan, Egypt	6670	3100
Mississippi	United States	6260	17750
Amazon	Brazil	6150	200000
Yang-Tse-Kiang	China	5800	35000
Yenisei	Russia, Mongolia	5540	20000
Obi	Russia	5410	13000
Huang-Ho	China	4830	20000
Amur	Mongolia, China, Russia	4667	11300
Congo	Democratic Republic of the Congo	4640	40000
Mackenzie	Canada	4600	15000
Paraná	Brazil, Paraguay, Argentina	4400	16000
Lena	Russia	4400	17000
Irtysh	China, Russia, Kazakhstan	4228	2800
Niger	Mali, Niger, Nigeria	4200	30000
Mekong	China, Vietnam	4180	50000

Table 7.2 - The five largest French rivers, classified by decreasing order of length

River name	Main tributaries	Source	Length [km]	Maximum and mean flow rate [$m^3 s^{-1}$]
Rhine	Aare, Ill, Neckar, Main, Moselle, Ruhr, Meuse, Ijssel	Saint-Gothard, Swiss Alps	1298	12000 1700
Loire	Allier, Cher, Indre, Vienne, Maine, Erdre, Sèvre Nantaise	Mont Gerbier de Jonc, France	1012	1830 930
Rhone	Arve, Ain, Saône, Isère, Drôme, Ardèche, Durance, Gard	Saint-Gothard, Swiss Alps	812	10000 1700
Seine	Aube, Yonne, Oing, Marne, Oise, Eure	Saint-Germain-Source-Seine, France	776	2000 300
Garonne	Neste, Ariège, Tarn, Lot, Dordogne	Val d'Aran, Pyrenees, Spain	647	1030 630

1.2. Uniform regime

Let us look first at a section of one of these rivers that is straight enough to be ana-lyzed like a straight, open canal. By watching for sufficient time the diverse objects floating on its surface, we find that the fastest flow is in the center, whereas the flow through some zones near the banks is much slower. Moreover, because the turbulent mixing near the free surface is efficient, the water speed is almost uni-form in the top part of the river, before decaying to zero at the bottom. Because of the low friction between the free surface and air, the maximum speed is sometimes located at an intermediate depth between the surface and halfway to the bottom. Measurements made in the field and in laboratory channels reveal both the mean speed profiles and the fluctuations with respect to this mean. Generally, the speed distribution looks something like that illustrated in figure 7.1. Because we are deal-ing with the turbulent regime, fluctuations in speed about the mean are found at each position, to the extent that the ratio between the effective fluctuations and the mean can be of the order of 3% to 10%. These fluctuations represent the contribution of eddies or whirlpools of all sizes, which are superimposed over and swept along by the average flow. Their signature is easily seen on the free surface and is due to the small differences in water depth associated with nonuniform distribution of pressure, which is generally a bit weaker at the center of whirlpools than at their periphery so as to compensate for the centrifugal force.[1]

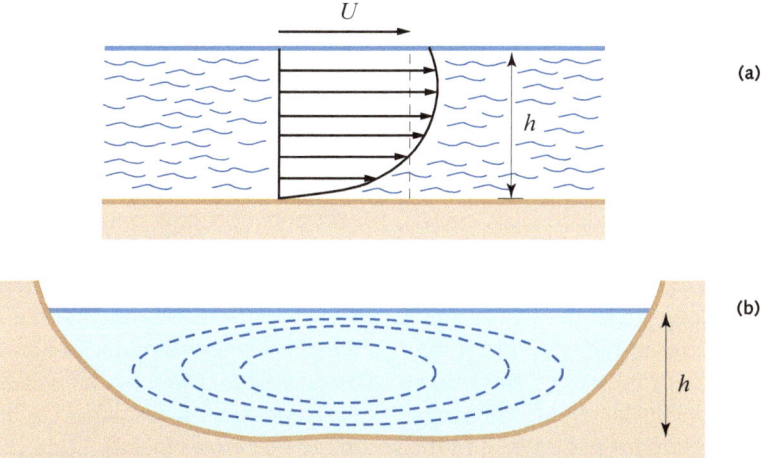

Figure 7.1 - Schematic view of the distribution of average speeds in a river cross section: **(a)** average flow speed U and the local distribution as a func-tion of depth, **(b)** isovelocity curves in a river cross section (dashed lines).

1   The pressure distribution in a flow is not determined only by the centrifugal force, even if this effect is dominant in the presence of rather-well-organized whirlpools. All local accelerations of the fluid are in general accompanied by a variation in pressure that is imposed by the condition of incom-pressibility. It is much more apparent in water than in air because it is proportional to the density.

In spite of this complexity, understanding the mechanisms that control the flow boils down, in a first approximation, to ignoring the nonuniform characteristics of the velocity distribution and the fluctuations. Let us assume that the actual velocity distribution shown in figure 7.1 can be replaced by a uniform distribution where U is everywhere the same in the cross section and that the depth is constant at h. This idealized regime is called the *uniform regime*. We shall focus on how the two parameters U and h are determined for any abscissa in the uniform regime, which corresponds fairly well to a rectilinear canal of constant width and slope. For this, we will exploit two laws that all longitudinal sections of this unit-length canal must obey. To make things as simple as possible, consider also a transverse slice of unit length, which equates to mentally isolating a volume of fluid of height h and of unit length and width.

If we neglect the very small amount of water lost to evaporation or that percolates into the soil, which is a very good approximation, the principle of conservation of mass implies that, because water is almost incompressible, the flow rate entering through the upstream face of this volume must equal the flow rate exiting the downstream face; in other words, the product Uh must not vary across the river. In addition, the driving force, which is a function of the bottom slope[2] of the river because it is proportional to the projection of gravity onto the bottom of the river, must be balanced by friction along the bottom. The combination of these two conditions allows us to determine the speed U and the depth h. Because the regime is turbulent (see appendix), friction is proportional to the square of the speed: U^2. The equation for the uniform regime is thus quadratic and so has two distinct roots. This property has the important consequence that two possibilities exist to convey the flow: one that is slow and deep and the other that is fast and shallow.

The regime is called *fluvial* in the first case because it corresponds to rivers that are rather calm, such as downstream parts of the Amazon or the Loire. It is called *torrential* in the second case, such as for torrents that flow down steeper slopes. Of the two possible speeds, one is systematically larger and the other smaller than the speed of gravity waves, which is $\sqrt{gh}$ when the depth h is small with respect to the wavelength λ of the waves. In chapter 6, in contrast, we saw that the speed of gravity waves in the open sea is $\sqrt{g\lambda/2\pi}$ when the depth is much greater than the wavelength of the waves; this result is also obtained for the speed of waves in deep water in the fluvial regime. In practice, we can consider the first speed to be a good approximation for $h < 0.05\lambda$ and the second to be better when $h > 0.5\lambda$. Note also that $\sqrt{2gh}$, which is slightly greater than $\sqrt{gh}$, is the maximum speed of a mass, initially at rest, undergoing free fall after being dropped in a vacuum from a height h.

---

2   In river hydraulics, the *hydraulic head* is usually the name given to the sum of three heights: the altitude z of the bottom of the canal, the depth h of the water, and the height $U^2/2g$. The hydraulic head, often denoted by H, can be interpreted as the energy per unit weight of the fluid. The line of the hydraulic head is the curve $H(x)$, not to be confused with the bottom line, which is the curve $z(x)$ and the water line, which is the curve $h(x)$.

1.3. Nonuniform regimes

Up to now, we have limited ourselves to the particular ideally simplified case of a canal with constant slope. Reality is always more complex, even in the case of a rectilinear canal because the variations in the slope must be taken into account all along the canal. This brings us to distinguish between two relatively simple yet opposite extremes: that of the regime where variations are slow, which hydrologists call *gradually varying* because the slope varies gradually along the abscissa (where the abscissa in this case and from now on represents the distance along the course of the waterway and is denoted x), and the regime where variations are very rapid, such as when the water breaches a weir or a spillway.

Let us begin by examining how the gradually varying regime differs from the uniform regime. The difference between the driving force and friction is no longer zero; rather, it is proportional to the acceleration of the slice of water, which is expressed by the derivative of its speed, or its depth, with respect to the abscissa. The equation [3] already evoked and that simplifies in the uniform regime to a second-order algebraic equation, now becomes a differential equation. It possesses remarkable properties, which translate the surprising diversity of flows that can be seen along a river whose bed has a gradually varying slope $i(x)$. The variation in average pressure along the cross section again leads to a term proportional to U^2, which comes from BERNOULLI's equation. The celerity c of gravity waves is a critical speed, so that, if we obtain $U = c$ at a certain point on the abscissa, the local slope of the free surface would become infinite, which contradicts the hypothesis of the gradually varying regime. This situation can only belong to the opposite extreme of the rapidly varying regime and must be excluded from our thinking at present. But we retain that this case corresponds to very specific values for the speed and depth, which are generally called the *critical speed* and *critical depth*. By letting $i = f$ in the numerator of the equation given in the footnote, we obtain $dh/dx = 0$ and find again the simplified equation of the uniform regime. We can see that, depending on the sign of the numerator $i - f$ and of the denominator $c^2 - U^2$, the depth can increase or decrease. This opens diverse possibilities for the shape of the water line; in other words, for the form of the curve $h(x)$.

In the fluvial regime, all sorts of perturbations can propagate upstream of their source with an effective speed of $c - U$. This difference is indeed the relative speed of the waves with respect to the free surface because the celerity c is the speed at which waves propagate relative to the ambient air, and the slice of water itself moves with speed U with respect to the ground. This relative speed is positive in the fluvial regime where $U < c$. In contrast, in the torrential regime, where $U > c$,

---

3   In the gradually varying regime, the equations for flow at a free surface reduce to the relatively simple form $dh/dx = [gh(i - f)]/(c^2 - U^2)$, where i is the slope of the bottom line and f is a dimensionless friction coefficient that is proportional to the square of the speed. The difference $i - f$ represents the dimensionless difference between the driving force and friction.

the perturbations can no longer move against the current. This may seem paradoxical for the observer that has often remarked that the form of the free surface of torrents is very agitated, whereas that of fluvial rivers is very calm. In reality, the agitation of the surface of torrents in the mountains is due to the tortured geometry of their beds formed of large stones, with immediate repercussions at the free surface as soon as the depth is of the same order of magnitude as the size of the stones, or smaller: each stone acts as a weir. In contrast, in a river of calm water, where the depth exceeds the critical depth, an irregular bed leaves almost no trace at the surface. For a large rock to affect the surface, its size would have to be of the same order of magnitude or greater than the critical depth h_c. If the height of this large rock is more than h_c, as is the case for artificial weirs, it causes a sudden change of regime. The flow becomes locally torrential above the weir and immediately downstream, and it recovers its fluvial characteristic a short distance farther downstream, after a sort of singularity called a *hydraulic jump*, which we shall examine more closely in a later section.

Artificial weirs are commonly used by the authorities responsible for maintaining the waterways or for managing the water resources. They introduce a loss of energy in the form of strong swirls that reduce erosion along the banks and localize it instead near the weir, which is made of concrete and regularly maintained. Upstream of the weir, they also facilitate the settling of suspended matter swept along by the river. In addition, they serve to measure and predict early on the flow that will arrive later into a downstream reservoir. Figure 7.2 illustrates the characteristics of a small-scale weir designed to locally dissipate a significant fraction of the stream's energy, so as to protect the banks. To measure the flow rate, much shorter weirs are generally used, and the height of the thin layer of water that spills over the edge of the weir is measured by contact. The flow rate is deduced after an a priori calibration is done in the laboratory on a scale model of the weir.

Several times we have evoked laboratory experiments made under controlled conditions and protected from the hazards of nature. These make it possible to precisely measure and test the influence of water works such as weirs, piles, bridges or the entrances and exits of locks. To guarantee similar results in the laboratory as would be obtained in the field, a simple rule must be followed: in each case the ratio of the various lengths (i.e., geometric similarity) and of the FROUDE number,[4] which is the ratio of the flow speed U of water to the celerity $\sqrt{gh}$ of gravity waves, is held constant. When this condition is satisfied,[5] the ratio between the potential

---

4   The FROUDE number, which is given by $Fr = U/\sqrt{gh}$, is proportional to the ratio of the average speed U in a river to the speed of a packet of water after free falling through a height equal to the depth h. It is named in honor of the British naval architect and engineer, William FROUDE (1810–1879).

5   To be absolutely rigorous, the theory of similarity demands that the REYNOLDS number also be identical in the laboratory and in the field. But, in fact, it suffices for the REYNOLDS number to be very much greater than unity in each case for this second condition to hold, as if it were infinity in both cases. This condition is thoroughly satisfied by a liquid such as water whose viscosity is very

energy (ρgh) and the kinetic energy ($\rho U^2/2$) exchanged per unit volume of water is the same in the laboratory as in the field. In fact, we can show that, if these ratios are held constant, the similarity extends to all mechanical quantities. Thus, all dimensionless ratios between similar quantities, such as speeds, flow rates, depths, pressures, or forces, will be the same in the laboratory as in the field. The theory of similarity thus leads to a powerful method that is often applied in hydraulics and in aerodynamics, because it allows us to use scale models to control the flow parameters, to make precise measurements, and determine the size of the setup. Ship hulls, sails, bridges, dams, turbines,[6] and pumps are all classic examples of optimization done by applying this approach. In aerodynamics, the wings of airplanes or the blades of helicopters can also be optimized by using laboratory scale models.

Figure 7.2 - Flow over a weir in the stream Sonnant d'Uriage near Grenoble, France. Note the uniform transition upstream from the fluvial regime to the torrential regime, then the cascade, then finally the transition from the torrential regime back to the fluvial regime after a hydraulic jump near the feet of the cascade. Note also the intense swirls, which are the signature of a strong dissipation of energy localized in the hydraulic jump. [© Philippe Belleudy, Université Grenoble Alpes, IGE]

Often in nature we find rivers in which the bottom slope changes suddenly from relatively steep, so that the regime is torrential, to very weak, so that the fluvial regime takes over. Upstream of the change in slope the depth slowly increases from an initial value that is less than the critical depth h_c, which is imposed by

---

low. This rule of similarity can be deduced from the Vaschy–Buckingham theory, which applies to all physical phenomena.

6   An example that shows the confidence accorded to this theory of similarity is the deals between builders and buyers of heavy equipment such as large turbines, who index the final sales price on the performance measured not on the real turbine, but on a scale model with an external diameter of 50 cm.

the upstream conditions (a weir, for example, or even the opening of a valve). Downstream of this change in slope, the depth also grows, until it reaches its limiting value far downstream, which is imposed by the downstream conditions and is greater than the critical depth (for example, the level of water in a lake or ocean into which this river flows). But the equation for the gradually varying regime, which is given in footnote 3, cannot connect the two laws for the water line $h(x)$, because its simplified form is not valid near the critical conditions; it also cannot determine the point where the critical conditions occur.

Figure 7.3 shows schematically what can be observed: near the change in slope but slightly downstream, the depth changes suddenly from a torrential value (less than h_c) to a fluvial value (greater than h_c). This quasi-discontinuity, called a *hydraulic jump*, is always accompanied by a strong dissipation of energy, as demonstrated by the large swirls found in these areas.

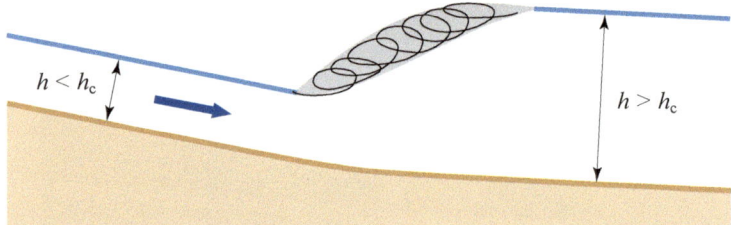

Figure 7.3 - Schematic view of a hydraulic jump in a canal, with the torrential regime upstream and the fluvial regime downstream. The gray region represents the swirls that dissipate a large part of the energy. The length of these swirls is often of the same order of magnitude as the depth. The blue arrow indicates the direction of flow. The slopes of the free surface upstream and downstream of the jump, as well as that of the bottom line, are so weak that, on the scale of the jump, these lines can seem horizontal.

Without even leaving home, each of us can observe an example of a hydraulic jump on a daily basis. Simply observe the almost circular jump executed by the water dropping from a faucet into the kitchen sink, as shown in figure 7.4. Note that the real length over which the jump spreads, which is made obvious by the accompanying swirls, remains fixed. It is the same order of magnitude as the depth in the fluvial regime; in other words, it is very small with respect to the other lengths characteristic of the water line and characterized by the function $h(x)$. In practice, this length is so short compared with the typical lengths of the flow upstream and downstream, themselves slowly varying, that we can treat it as a discontinuity for a theoretical analysis.

The rapid transition from a torrential regime to a fluvial regime is completely analogous to the transition from a supersonic regime to a subsonic regime in an aerodynamic nozzle or around an airplane (see chap. 4, section 4.3). The water depth is analogous to the density of air, gravity waves are analogous to sound waves, and the hydraulic jump is analogous to the shock wave. In both cases, the critical conditions correspond to equating the speed of the fluid with the celerity. Waves cannot pass upstream from the fluvial regime to get to the torrential or supersonic regime because the flow in these latter regimes is faster, and the hydraulic jump, like the shock wave, is accompanied by a significant dissipation of energy.

Figure 7.4 - A nearly circular hydraulic jump in a sink. Its flow is torrential between the foot of the stream of water and the hydraulic jump and is fluvial at the exterior of the hydraulic jump. The flow rate per unit area is constant over the entire surface centered on the foot of the stream, with the same flow rate exiting through the drain. The speed is thus much greater in the torrential zone than in the fluvial zone. The presence of swirls near the jump is indicative of the significant energy dissipation in this area. Note that the instabilities of the free surface, which are everywhere in the fluvial zone, do not manage to move upstream through the torrential flow. [© Philippe BELLEUDY, Université Grenoble Alpes, IGE]

The symmetrically opposite situation also occurs in which the slope of the river bed goes from a low value upstream, betrayed by the presence of a relatively calm basin, to a high value downstream, immediately associated with a torrential regime. But this case is less spectacular because the transition is uniform and does not have the singularity of the hydraulic jump. Also, in the case shown in figure 7.2, we can observe both the uniform transition *fluvial* → *torrential* immediately upstream of the weir, with no significant dissipation of energy, and the critical transition *torrential* → *fluvial* downstream of the waterfall, accompanied by a strong dissipation of energy in the swirls. Even if the equation for the water line for the gradually varying regime is no longer valid close to the jump, it does announce the singularity that corresponds to the vanishing of the denominator ($c^2 - U^2 = 0$) without a proportional vanishing of the numerator. It also reveals the existence of the regular transition when the numerator ($i - f$) and denominator ($c^2 - U^2$) both go to zero at the same point.

A rather particular form of hydraulic jump, called a *tidal bore*, may be seen during an incoming tide in estuaries subjected to large tidal flows. While the tide lifts the water level downstream, a wave traveling up the estuary encounters water that becomes progressively less deep, often accompanied by a decrease in the width of the river bed. The slope of the upstream face of the wave thus increases until, when it becomes almost vertical, it breaks, sending water upstream and announcing the arrival of the tide (fig. 7.5). The waveform, with a very sloped face that becomes almost vertical before breaking, and a downstream face that becomes less and less sloped is also encountered at sea near beaches and was already discussed in chapter 6.

Figure 7.5 - Tidal bore at Vayres on the Gironde estuary, France.
[© José Arnal, President of the association *Le Mascaret* at Vayres]

A sudden variation in the width of the river bed generates effects that are analogous to those caused by a change in slope. This effect is visible, in particular, around bridge pilings, or when a lock is being filled. When bridge pilings are sufficiently large and closely spaced so as to significantly restrict the flow of water through them, the average speed increases locally so as to conserve the flow rate. In so doing, they provoke a reduction in depth, which can become sufficient to change the flow regime from fluvial to torrential, at least very locally between the bridge pilings, before returning via a hydraulic jump to a fluvial regime downstream of the bridge (fig. 7.6). In this case, however, because the flow becomes converging, the approximation of parallel flow lines no longer holds, and the calculation of the free surface requires a more refined theory than that presented above.

Figure 7.6 - The DUMNACUS bridge at Ponts-de-Cé on the Loire, near Angers, France. The fluvial regime is visible to the left through the first arch and transforms into a torrential regime under the arch, where the speed increases to compensate for the narrowing of the water channel between the large bridge pilings. In turn, several meters downstream, this torrential regime transfers back to the fluvial regime via complex-shaped hydraulic jumps, accompanied by swirls and white emulsions. [© Les Ponts-de-Cé]

2. Curves and meanders

The preceding paragraphs allowed us to better understand the diverse forms of free surfaces that can be observed along rectilinear waterways, where the flow is parallel to the central axis. But in general, and especially in plains where the slope is small and the flow regime can only be fluvial, nature builds waterways more or less curved that can evolve to the point of meandering. Historical records show that the amplitude of these meanders grew larger and larger over the millennia until, to limit their extent and preserve arable lands and the communities on both sides of the river, in some countries levees were constructed, regularly maintained, and planted with vegetation capable of solidifying the banks. What mechanism is behind these meandering forms and their progressive growth?

The new element is the curving central axis of the waterway, which leads to a centrifugal force that pushes the water toward the outer bank (fig. 7.7). On average, this force is balanced by the pressure difference that builds up so as to be greater on the outer bank than on the inner bank. This pressure difference is accompanied by a difference in the level of the free surface near the two banks. Even if it is hard to see by eye, this height difference is present and can be measured. However, the centrifugal force, which is proportional to the square of the local water speed, is not distributed uniformly across the river cross section. We must differentiate between the zones near the free surface, where the speed and the centrifugal force are near their maximum, and the zones nearer the bottom where these quantities are significantly less. Near the surface, the average equilibrium is tipped in favor of the centrifugal force, which deviates the current toward the outer bank. The current can thus progressively erode and excavate this bank (if it is a natural bank) and thereby weaken the terrain. Erosion also explains the presence of exposed tree roots on this bank, because the earth that covered the roots has been progressively carried away. Near the river bottom, the situation is the opposite: the equilibrium is tipped toward the centripetal pressure, which guides water toward the inner bank.

Figure 7.7 - Schematic representation of eddy generated by the centrifugal force in a river bend.

This interplay between forces creates an eddy in the cross section of the river shown in figure 7.7, which is superposed over the main current. Its upper part, close to the surface, is directed toward the outer bank, whereas its lower part, near the river bottom, is directed toward the inner bank. The load of earth, sand, and stones produced by the erosion of the outer bank is thus brought toward the inner bank where the water speed, which is generally fairly low, allows it to be decanted and deposited. Anyone can observe that the small beaches that appear along rivers, where the depth increases very slowly, are located on the inner banks of the meanders whereas, on the contrary, the water rapidly becomes very deep on the outer bank, which is the bank chosen by divers. In the language of physics, this form is called *unstable* because the amplitude of the meander is constantly growing, even if the excavation of the outer bank is extremely slow. This instability is what, century upon century, has forced numerous rivers to dig their meanders through the plains that they irrigate, and that they can also destroy if man does not intervene to limit the amplitude of the phenomenon. The example of the meander in Queuille in Auvergne, France (fig. 7.8) shows that, given time, even a river such as the Sioule is capable of digging significant meanders, to the point of almost closing a loop around the peninsula, which could become an island if protective infrastructure is not periodically applied.

Figure 7.8 - The meander near Queuille on the Sioule in
Auvergne, France. [© Yannick MASSON, Mayor of Queuille]

A comment is in order here to better understand the significant difference between the meanders of rivers with very gentle slopes, such as the Seine or the Mississippi, and those of rivers such as the Garonne and the Isère that descend from the Pyrenees and the Alps in much steeper valleys. These latter rivers transport a large load of debris consisting of massive objects: rocks of all sizes or branches and trunks of trees ripped from the banks or from torrential tributaries farther upstream. These waters have a large erosive power, both because their speed is greater due to the steep slope of their bed, and because the large pieces of debris serve as battering rams whose constant hammering efficiently dig away at the natural banks, even if these latter have strong cohesion. Such debris can also damage man-made levees. The outer banks of mountain rivers must thus be constantly monitored, notably after flood levels are reached, and periodically restored by civil engineering works.

3. Waterfalls and cascades

Numerous waterways offer the spectacle of superb waterfalls, which can attract hordes of tourists. The most notorious example is, without doubt, the Niagara Falls in North America on the Saint Laurence River (fig. 7.9), which attracts over 10 million visitors annually. Such a large number of visitors has led to the development of the Niagara Parks Commission, with its hotels, restaurants, expositions, and transportation network. But the small cascades high in the alpine valleys also attract numerous hikers each year. All these waterfalls, large or small, oblige the local authorities to control the flux of tourists and often to undertake serious public works. Apart from the beauty of these sites, illuminated as they are by the stunning colors of the waters, what noteworthy properties do these waterfalls possess?

Figure 7.9 - Panoramic photograph of the Niagara Falls. [© The Niagara Parks Commission]

From the point of view of hydraulics, the most important property is certainly the high speed attained by each layer of water in quasi free fall. We have already seen that the speed of an object free falling in a vacuum through a height h is given by $\sqrt{2gh}$.[7] For a 100 m fall, this is of the order of 45 m s^{-1}; for a fall of 10 m it is of the order of 14 m s^{-1}. Seen by a nearby tourist, these speeds (160 km h^{-1} and 50 km h^{-1}, respectively) seem staggering. In reality, because of the instabilities of

---

7   The speed of an object in free fall in a vacuum is the result of the transformation of its potential energy ρgh per unit volume into kinetic energy $\rho U^2/2$. The density ρ does not enter the expression for the speed of free fall, which is thus the same for a feather as for a lead brick. Yet a feather and a lead brick do not fall at the same speed! This famous puzzle is generally attributed to TORRICELLI, GALILEO's student, who was the first to do this experiment in a tube with sufficient vacuum. The enigma is easily resolved: the feather and the brick do not undergo free fall but are restrained by air resistance, which is much more efficient in slowing the feather than the brick.

the free surface, the friction between water and air is not as negligible as one might think. The free surface takes on a very complex form, from which numerous droplets are torn off and, with them, a significant momentum.[8] The free surface is nothing like a smooth wall but is more like a rough wall. The speed is thus less than that given above, although the orders of magnitude are correct.

The consequence of these high speeds is the destructive power and the boring capacity of these waterfalls, even when they strike a solid surface, such as a rock. This explains why waterfalls often land in cuvettes that they have hollowed out themselves over the centuries. However, the boring power of a waterfall is notice-ably reduced when it falls into a rather deep cuvette that contains enough water to absorb the impact of the waterfall and dissipate its energy. This mechanism explains why the boring rate decreases over time until it finally stops altogether.

The other remarkable property of waterfalls is the frequent presence of fog around the falls and the cuvettes that receive them, as is easily seen in figure 7.9. This cloud, which practically sits on the water in the cuvette at the base of the falls, has a double origin. For starters, it contains the remains of droplets after they are torn from the falls and partially evaporate depending on the capacity of the surrounding air to dissolve water vapor; in other words, the saturation vapor density. But we also see a cloud that constantly reforms above the cuvette, which shoots a signif-icant fraction of the flow it receives back up into the air in the form of a spout of water. The height of this spout can be several meters, after which it falls back into the cuvette to contribute strongly to the mixing therein. Air and water are thus strongly mixed, and the droplets formed are rapidly renewed before they can evap-orate. Around the Niagara Falls, this cloud is called *mist* and one of the boats used to bring tourists, protected in their single-use waterproof liners, near the spout is called *Maid of the Mist*. With the air full of droplets and cloud, beautiful rainbows may be seen on nice days, and not only at the Niagara Falls, but also around the spouts at the feet of almost all large waterfalls.

Conclusion

The goal of this chapter is to introduce the reader to the most notable phenom-ena that occur along waterways, and to describe rather precisely the mechanisms behind them. The most directly observable curiosities of flowing water were emphasized: water depth, speed, hydraulic jump, excavation of meanders, and spectacular waterfalls. Numerous other aspects that nature conceals from strollers might also have merited attention, such as the infiltration of water into the ground and the exchange with the water tables. But the associated flow rates are much

---

8   It can be rigorously shown that the momentum exchange due to the movement between two neighboring fluid layers is mechanically equivalent to friction. This holds whether the two fluids are the same or different; the latter is the case of a liquid cascade surrounded by air.

less than the main currents; the interested reader can find detailed information on these subjects in specialized texts.

This chapter would not be complete without a word of warning about the dangers of all these waterways that attract tourists and along which tragic accidents sometimes occur. Descending a river in canoe by fine weather may delight the enthusiasts, but can also become catastrophic during a violent, unexpected flood that can follow on the heels of a thunderstorm localized over a tributary. Large rivers are less whimsical because of their inertia and the infrastructure that has already been built around them. In fact, many of them can now be navigated. But one is well advised to be wary of the convulsions of these sleepy giants.

Chapter 8

Lakes, dams, and major works

On waves beneath the skies, afar and wide,
Naught but the rowers' rhythmic oars we heard
Stroking your tuneful tide.

(Alphonse DE LAMARTINE, *The Lake*)

© Springer International Publishing AG 2017
R. Moreau, *Air and Water*,
DOI 10.1007/978-3-319-65215-3_8

A walk along a river often leads to a quiet lake or pond, be it natural or artificial. Present at all altitudes, these lakes, ponds, or marshes are all mirrors of the surrounding nature. While they may participate in forming the beauty of the scenery, they also fulfill important functions in ecology, energy, industry, agriculture, and tourism. Some formed simply when streams and precipitation filled natural basins left behind by retreating glaciers, or in the crater of an ancient volcano. The free surface rose to the level that allowed flow into the lake to be evacuated by the effluent, as well as by evaporation and percolation through the soil toward the water table. Other reservoirs, made by man, testify to his efforts to store the immense quantities of freshwater necessary for the demands of agriculture and energy. Along rivers, notably the most whimsical, modern man has managed to create infrastructure that serves multiple purposes: spreading out floodwaters, irrigating agricultural lands, and producing energy. And along the coasts, which must be protected from the assaults of the sea, civilization reveals its desire to constrain nature by constructing the ports and piers required by sailors and multiple other, sometimes grandiose, works. Thus, with the stupefying acceleration linked to the industrial development of the 20th century, the technology of large hydraulic works has developed. The sight of these spectacular works and the knowledge of their origin, their role, and their functioning, impresses upon us just how vital water is for humanity and how important its control is for modern society.

1. From marshes to hydroelectric reservoirs

Let us begin with the ponds and marshes, which generally are natural, and their ecological role. To cite only several examples from France, consider first the thousands of ponds near La Dombes, which cover about 10 000 hectares in the department of Ain between Lyon and Bourg-en-Bresse. Their fishery resources nourish numerous species of birds and attract each year large migratory populations. The ecosystem of this region, where each predator finds its prey and where the hunting and fishing contribute to the local gastronomy, seems to have succeeded in finding its balance. Close to the Atlantic Ocean, the Poitevin marsh, where trails and canals tangle in a network so complex that a map and compass, or even a guide, prove most useful, has its own particular and unique character. This marsh is the result of centuries of human activity. It is in fact man that dug all the shallow canals and backfilled all the paths by using material pulled from the canals, with the twin goal of allowing animal species to find habitat and food, and to ensure for himself a constant supply of game and fish. The ecological balance of this marsh seems less sure than that of the marshes near La Dombes because it is beginning to feel the effects of the intensive agriculture that has developed around its periphery. Another example from France is that of the salt marshes of Camargue, in the Rhone delta near the Mediterranean Sea. Its ornithological park, it wildlife, and its ancient traditions that center on the manades, horses, and bulls makes this a special place to discover a relatively protected natural site that is, however, still subject to the whims of the weather that sometimes lead to invasions by the sea.

Along the rivers and streams, levees and dams create reservoirs that have diverse goals, such as to regulate the flow by spreading out floodwaters, while withdrawing part for irrigating nearby agricultural lands or for running turbines to produce electricity. Beginning at the opening of the 19th century, numerous rather home-made installations were built in the valleys of mountainous regions, one after the other, to capture the energy of the torrents. At appropriate places, a gently sloping canal, often called a *mill race*, would deviate a fraction of the flow to bring it to a narrow moderately sloped canal several meters long where a wooden water wheel was installed. The flowing water rotated the water wheel, whose axis was connected to machines by gears or belts. These machines replaced human or animal force to carry out the essential functions of these times, such as sawing and processing wood, which was an important resource of the mountain regions.

Bit by bit, as industry began to require significant energy resources, larger and better-engineered works were conceived and installed. In France, the chemical industry was the first to construct their factories in the mountain valleys, near waterfalls several tens of meters or more in height, which required preliminary studies and significant works to tame. Similar works were done in almost all the mountains ranges in France and in the other counties of Europe and North America. Next, as soon as it was industrialized in the 20th century, the electricity fairy invited herself into the process of transforming energy. The turbines and alternators were turned by hydraulics, producing electricity that was distributed over a local network, supplying the electrochemical industries installed nearby, driving the machines in the workshops and lighting the cities and villages. By using the hydrological resources of the mountain valleys, this industrial organization attracted and maintained a local population that was much greater than what could have been supported by agriculture alone.

By separating the use of energy from its place of production, the spread of electrical energy in the second half of the 20th century had very significant consequences. Instead of associating a small stream with a small power plant built nearby, hydroelectricity allowed powerful falls to be harnessed to complement coal power; these plants were then connected to a national grid to distribute energy over the entire national territory, regardless of where it was produced. The economic necessities of the end of the 20th century contributed to this evolution by demanding ever larger power plants, many built near large urban centers where manpower was available, or near ports or large transport links. This development immediately condemned the old, small, and relatively inefficient power plants of the mountain regions. Many of them were forced to stop operations and their buildings are now gutted by ruin, if not already demolished. The landscape is profoundly marked by this evolution, which forced the local population to rapidly adapt or suffer dire consequences, although some surprising repercussions ensued. The end of operations at the aluminum electrolytic production plants at Rioupéroux in the valley of the Romanche near the town of Livet-et-Gavet in Isère, or at Tarascon in the valley of the Ariège, or again at Argentière-la-Bessée in the valley of the Durance, in conjunction with the development of tourism and winter sports in these mountain regions, illustrates these transitions and rebounds.

The end of the 20th century was marked by a new evolution that associated nuclear power plants, very powerful but difficult to tune to follow the fluctuations in demand, with hydroelectric plants, which adapt rapidly to demand. The hydroelectric system, which requires two reservoirs not too far apart and at different altitudes, was thoroughly perfected by its previous use in the industrialized mountain valleys. During peak electricity demand, the gates of the upper reservoir are opened, water flows down to the lower reservoir where it is stocked and, in so doing, traverses a hydroelectric plant. This plant, equipped with turbines, alternators, and transformers, produces electric power that is sent over the national grid where it is added to the power from nuclear plants. Thus, during this time of peak demand, the upper reservoir loses part of its initial stock of water. In contrast, at night, when electricity demand weakens and the nuclear power plants continue to produce electricity, the reversible machines of the hydroelectric plant use the available excess energy to lift the water from the lower reservoir back to the upper reservoir. The capacity of the upper reservoir, which was conceived to stock water on the scales of the seasons, is much bigger than the lower reservoir, whose volume needs only supply the nocturnal pumping.

Two technological advances were required to obtain this performance. First, the hydraulic machines, called *reversible turbines*, had to be perfected to function in both directions with acceptable efficiency. Moreover, a system had to be set up to manage this hydraulic resource on large scales—at least on the national scale—to constantly optimize this water resource and to provide crucial support for the decision makers: which gate to open or close each second to keep instantaneous supply equal to demand without wasting any precious water? The narrow connections between electric-power distribution networks on a European scale make managing the hydraulic resources a particularly demanding task. Computer-aided decision tools have thus acquired considerable importance.

As for the landscape, it now displays two reservoirs instead of one. One of the most-cited prototypes of this type of hydraulic facility, coupled with nuclear power plants by the distribution grid, is the Grand'Maison dam in the town of Vaujany along the Eau d'Olle, which is a tributary of the Romanche in Isère, France. The upper Grand'Maison reservoir, which can hold 140 million m^3 of water, is at an altitude of 1590 m. It is associated with a lower reservoir of much smaller capacity, generally called a *discharge basin*, which can hold 15 million m^3. This discharge basin is located in the vicinity of Le Verney, 900 m lower and immediately adjacent to the hydroelectric plant, which is equipped with eight reversible turbines and four PELTON turbines.[1] Together, these machines generate 1820 MW of power. Figure 8.1 shows part of the Grand'Maison reservoir with the embankment, which is sufficiently inclined for a road to switchback up its face. This construction of earth and rock fill, built between 1978 and 1985, was filled in 1988. It belongs to the category of gravity dams, named as such because their stability is due to their weight, which can resist the pressure of the water on the upstream face of the dam and prevent it from tilting below dam.

Figure 8.1 - Grand'Maison earth and rock-fill dam on the Eau d'Olle in Isère. Length is 550 m, height is 140 m, width at base is 520 m, width at top is 10 m, volume and surface area of reservoir are 170 million m^3 and 2.2 km^2, available power is 1820 MW. Note the switchback road on the fairly gently sloped face of the dam. [© Franck ODDOUX – EDF/PWP]

Note that pumping water from the lower reservoir to the upper reservoir consumes more energy than the inverse operation produces. At Grand'Maison, the pumping consumes about 1700 GWh each year, whereas the turbines produce about 1400 GWh. The difference, which is close to 21%, is compensated by the fact that the dam makes the nuclear power plants more profitable by using the energy they produce during off-peak hours. In other words, this loss of energy leads to an overall gain.

Let us return now to reversible turbines to clarify how their design differs from that of conventional turbines, which are described in section 4 of this chapter. To a central wheel are fixed blades that receive the force of the water in turbine mode or push the water in pump mode, with the wheel itself sandwiched between disks positioned both above and below (fig. 8.2). These disks, which do not figure on propeller turbines, play a double role. First, they force the water to follow a channel

---

1   These turbines (see fig. 8.10) were invented by the American engineer Lester Allan PELTON in 1879.

so as to optimize the efficiency of the wheel by minimizing recirculation or second-ary vortices, which lead to significant energy losses. Second, they allow several wheels to be stacked on a single axis one above the other, separated by these disks (this is not the case for the single-stage turbine shown in figure 8.2). With these disks, multi-stage reversible turbines become possible; these are suitable for very fast flowing water, which is spread out over the successive stages.

Figure 8.2 - Single-stage reversible tur-bine, which is capable of functioning in both directions with water drops between 60 and 700 m. In turbine mode, the water comes from the periphery and crosses a series of directional blades of variable ori-entation (bronze color) and flows over the wheel. The vertical axis powers an electric machine situated above the wheel (not shown) that operates as an alternator and produces electric power. In pump mode, on the contrary, the electric machine functions as a motor and makes the wheel turn, which pushes the water back up to the upper res-ervoir. [© 2017 General Electric Company]

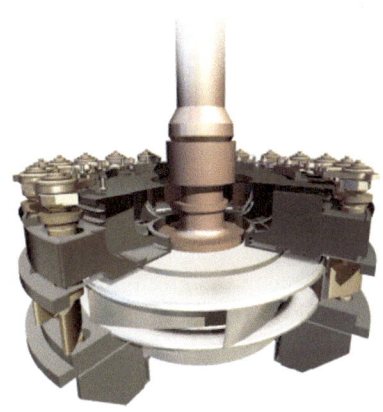

The power provided by nuclear power plants (a typical one consists of several stages, each of which provides about 1200 to 1500 MW) oblige them to be located near waterways with large flow rates, such as the Rhone for the Bugey or Tricastin reactors. The water from the river serves as the cold source for the overall thermodynamic cycle.[2] A fraction of its flow is diverted and constitutes the heat-transport fluid in a series of three circuits, the last of which connects to the exterior. In French pressurized-water reactors, the primary circuit circulates around a closed loop at high temperature (about 300 °C) and, to avoid boiling, at high pressure (about 1.5×10^7 Pa). The water withdraws heat from the reactor core and transfers it through the containment structure surrounding the core to a secondary circuit,[3] itself at the appropriate temperature and pressure for operation of the turbine and of the condenser immediately following. The secondary circuit is also a closed loop and produces vapor that blows over the gas turbines to spin the alternators and produce electricity. Finally, in the third circuit, the heat that was not converted into

---

2   As happens in all facilities where energy is converted, a nuclear power plant follows a thermody-namic cycle from a heat source (the reactor core in this case) to a cold source (the river water). This cycle is governed by the CARNOT efficiency, which is given by the ratio $T_h - T_c/T_h$, where T_h and T_c are the absolute temperatures of the heat source and the cold source, respectively. This is the maximum possible overall efficiency of the system; no real machine can attain it.

3   Although the thermal energy is transferred by conduction through the containment wall of the reactor core, the slightly radioactive water in the primary circuit does not mix with the water of the secondary circuit, nor *a fortiori* with the water of the third circuit, which is constantly exchanged with river water.

electricity is evacuated to the environment. Part of this waste heat is sent into the atmosphere, generally by large cooling towers from which large steam plumes escape and condense as they cool. The remaining heat is sent back into the river, carried by water that is several degrees hotter than the upstream river water.

2. Large dams:
stability and conforming to site

At this point, let us sketch a brief panorama of the main types of hydraulic facilities that we may find along our wanderings and explain their form and how they are adapted to their sites. We have already evoked the case of gravity dams made of earth (in the discussion of Grand'Maison), but gravity dams may also be constructed from concrete. The stability of the structure is still due to its weight, which must be able to resist the water pressure without requiring particularly solid connections with the ground at its extremities. This structure is thus used in valley bottoms where the rock or soil is not particularly cohesive or resistant.

The largest dam in this category is the gigantic Three Gorges Dam in China on the Yang-Tse-Kiang River, which retains a body of water with a surface area of 1080 km^2 and whose hydroelectric plants are designed to deliver 18 200 MW of power, which represents 10% of the total power capacity of this huge country. The parameters announced for this project are given in the caption to figure 8.3. The construction of this dam, which was done in stages, started in 2006. At the time that these lines are being written, the date of its completion and whether it will attain its original goals remain uncertain. The sheer size of this project was cause for rather animated discussions, or even controversies. The extent of these can be appreciated from the fact that the decision to construct was approved (with 1767 votes in favor) by China's National People's Congress only with a record number of abstentions (664) and even votes against (177), which is very unusual in this country. Arguments in favor were linked to the quantity of energy to be produced in a China that aspires to economic development. Rather solid arguments against the project were also raised: flooding of 600 km^2 of inhabited and cultivated lands, displacement of millions of people very far from their original home, risk of sedimentation upstream of the dam, and upheaval of the immense downstream ecosystem, notably due to the drying of wetlands and swamps.

In the mountains, where valleys narrow between two rocky slopes that geological examinations have guaranteed to be solid, thin dams of impressive elegance have been constructed. One of the best known is the Chevril Dam at Tignes, in Savoy, France. The history of this large dam is marked both in the records created in 1952 when it was put in service (see caption of fig. 8.4) and by the lawsuit that preceded its construction. Initiated by the inhabitants of the old village of Tignes, which is now swallowed up by the waters, and by their supporters, this lawsuit resulted finally in the decision to construct the dam. Ever since, every ten years when the EDF (the French national electric company), who operates the site, empties the reservoir to clean the intake pipes, a pilgrimage is undertaken to the ancient village whose inhabitants have since been rehoused higher up in the village of Les Boisses. The bells of the old church, as well as the statues and the gilded leather that adorned them, were transferred to the new church. Before the arrival of this abundant energy, the original inhabitants made a living from livestock and crafts in a superb natural site but at the cost of particularly painstaking work. Their descendants chose careers in winter sports and tourism, thanks to the development of the Tignes ski area, which is one of the best known in Europe, especially since it hosted the Winter Olympic Games in 1992.

Figure 8.3 - Three Gorges Dam in China, which was filled in stages starting in 2006. Length is 2335 m, height is 100 m, surface area of retained water is 1080 km², power upon completion is 18 200 MW. [© Christoph FILNKÖßL]

In France, however, the dramatic failure of the Malpasset arch dam (fig. 8.5) remains in the collective memory and continues to cause anxiety among the inhabitants of valleys below dams. This catastrophe occurred in the evening of December 2, 1959. The circumstances were extremely unusual: heavy rains along the Mediterranean coast caused an unusual rise in the water level of the reservoir at a time when the gates could not be opened due to construction of the future highway downstream. The reservoir above the dam thus reached maximum capacity when, at 22h13, the rock on the left bank into which one extremity of the arch was anchored broke free and took with it three-quarters of the dam. A large gaping hole was created, allowing the 50 million m³ of water in the reservoir to surge down onto the villages and fields, all the way to the town of Fréjus, which was hit by the flood crest some twenty minutes later.

Figure 8.4 - Chevril arch dam at Tignes, on the Isère River, put in service in 1952. The arch is 180 m high, 295 m long, and 43 m thick. This reservoir supplies 805 GWh per year, which corresponds to the energy consumption of a city such as Grenoble. The painting on its downstream face of HERCULES holding up the arch is the work of Jean-Marie PEIRRET (1989); oxidation has since made this portrait of HERCULES more and more difficult to recognize. [© Franck ODDOUX – EDF/PWP]

The waters formed a giant wave some forty meters high that rushed headlong at 70 km h^{-1}, sweeping away everything in its path until being absorbed by the sea. From among the inhabitants of the valley, 423 victims were counted.

Figure 8.5 - Photographs of the upstream face of the Malpasset dam before it was filled and after its failure on December 2, 1959.

Thousands of dams have been built worldwide and, statistically, one or two fail per year. In general, these are smaller dams, located in sparsely populated areas, so their failure is not considered a catastrophe. The reasons behind these failures are almost always natural (earthquakes, floods, devastating cyclones). In fact, of the 40 failures documented from 1970 to 2001, only three do not have natural causes: the Morvi dam in India, which failed in 1979 after the water level rose excessively because of insufficient drainage by an undersized spillway; the Bhopal dam also in India, which failed in 1984 due to an explosion in a nearby chemical factory; and the Philippines dam, which failed after a collision between a ferry and an oil tanker in 1987. Despite the catastrophe of Malpasset, arch dams have proven to be the most solid. The accident at Vaiont in the Italian Alps, which happened four years after Malpasset, illustrates well their robustness. October 9, 1963, a whole section of the mountain collapsed into the reservoir, generating a 100-m-high wave that swept over the dam without destroying it and then continued down the valley, where it claimed 2600 victims. Specialists in materials resistance explain the robustness of these dams by invoking the water pressure on the arch, which subjects it to compressive forces, thereby increasing its strength. At the risk of oversimplification, recall that anyone who is exposed to a strong wind can more easily resist it by turning a round back to it rather than by facing it head-on.

In the category of large dams, whose list is regularly updated by the International Commission on Large Dams (ICOLD), we must also cite buttress dams, also called *multi-arch dams*. The technique of buttressing consists of holding back the water with a fairly flat wall, supported on its downstream face by reinforcements imbedded into the foundation. The wall spreads the pressure of the water partially onto the banks and partially onto the foundation. Each portion between neighboring buttresses is analogous to an arch dam, which justifies the term *multi-arch*. A buttress dam is used when the lateral anchoring points are too far apart to construct a single arch, or when the local material is too compact, making excavation too difficult to justify a gravity dam. Buttress dams also have the advantage of being very Spartan in terms of their material requirements.

One of the largest examples in this category is the DANIEL–JOHNSON Dam in Quebec (Canada), which was put in service in 1970. With a height of 214 m and a length of 1312 m, it is supported by two central buttresses separated by 160 m at their base and 13 lateral arches that form inclined semicylinders. Construction of this dam required over 22 million m^3 of concrete, which is about five times less than a gravity dam. In France, the most splendid dam of this type is no doubt the Roselend Dam (fig. 8.6), which was put in service in 1962 and which served as a prototype for the ambitious DANIEL–JOHNSON project.

Figure 8.6 - Roselend Dam in Savoy, France, put in service in 1962. It is 800 m long, 150 m high, and has a capacity of 185 million m^3. [© Franck ODDOUX – EDF/PWP]

3. Management of large rivers

Various large and heavily visited works have been built in the run of rivers, one upstream of the next, with the double goal of taming the overflows of impetuous rivers and of transforming their hydraulic energy into precious electrical energy. The management of the Rhone, which is done by the National Company of the Rhone (CNR), seems like a good example. It started in 1933 with the beginning of the Génissiat Dam downstream of Lake Geneva, which is a natural lake of glacial origin. The project was heavily perturbed by the Second World War, because the construction site was flooded on purpose for several years, and so was not completed until 1948. It is considered as the first large hydroelectric dam in France. Its characteristics, which are summarized in the caption for figure 8.7, show that its installed power of 420 MW was already quite respectable and justified the interest it attracted in the middle of the 20th century.

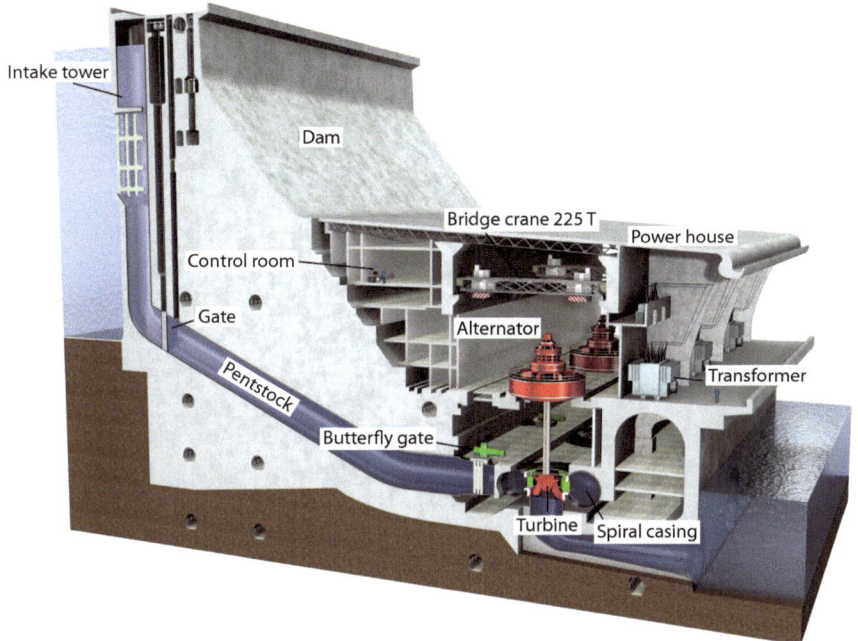

Figure 8.7 - Hydroelectric dam and power plant of Génissiat Dam. Cross section showing the main parts, which are integrated remarkably into a single structure. [© Compagnie nationale du Rhône]

This project had considerable impact: it demonstrated the feasibility of large dams and opened the way to the integrated management of the entire Rhone from where it leaves Lake Geneva until its mouth at the Mediterranean Sea. Potentially, with the altitude of Geneva being at 370 m and an average flow rate of 1700 m^3 s^{-1}, the power available at any time[4] along its French portion is in the gigawatt (GW) range.[5] This order of magnitude reveals both the potential hydroelectric resources of this river and the potential damage that could be caused by the largest floods. Apart from electricity production, managing the Rhone also served other purposes, which we shall return to later.

The other large and well-known dam built on the Rhone is the Donzère-Mondragon Dam, which is described in detail below. Let us clarify first that the Génissiat and Donzère-Mondragon dams are part of an impressive series of 18 hydroelectric power plants and 5 nuclear power plants that are installed along the French part of the Rhone (fig. 8.8). This powerful series of plants justifies well the consideration of the river as the most powerful energy deposit in France.

Figure 8.8 - The André BLONDEL hydroelectric plant bordered by European-sized locks near Bollène, at the extreme downstream end of the canal that starts at Donzère Dam on the Rhone. [© Compagnie nationale du Rhône]

The Donzère-Mondragon Dam is in fact one of the dams that allowed the down-stream part of this large river that is the Rhone to be canalized, just south of Lyon. Although narrow, the Rhone corridor has been transformed into a well-controlled and absolutely critical artery for the Europe of the 20th and 21st centuries. While

---

4 The power of a waterfall is given by the product of its flow rate Q and the difference in hydrostatic pressure associated with the drop h, which is itself given by the expression $\Delta p = \rho g h$ (ρ denotes the density of water, g is acceleration due to gravity). The power P of a waterfall is obtained from the formula $P = \rho g h Q$. Similarly, the power furnished by a pump that delivers a flow rate Q with a pressure difference Δp is given by the formula $P = \Delta p Q$.

5 1 GW (gigawatt) = 10^3 MW (megawatt) = 10^6 kW (kilowatt).

we watch the considerable flotilla of barges, trains, and trucks that travel daily along this corridor between the Mediterranean Sea and Northern Europe, can we still imagine what life was like in this valley during the 19th century when Alphonse DAUDET described and Paul CEZANNE painted their native Provence?

Although the installed power of the Donzère-Mondragon Dam, at 354 MW, is of the same order of magnitude as that of the Génissiat Dam, the annual capacity is slightly greater: more than 2100 GWh compared with 1700 GWh for the Génissiat. The difference is small because the two main contributing factors vary inversely: a much larger flow rate of nearly 2000 m^3 s^{-1} at Donzère after the significant contributions of tributaries such as the Ain, Saone, and Isere rivers versus 750 m^3 s^{-1} at Génissiat; but a drop of 67 m at Génissiat versus 23 m at Donzère-Mondragon. Apart from electricity production, management of the Rhone has other important objectives, such as protecting the valley from floods, irrigating the arable land, and large-scale shipping, especially for merchandise.

This aspect is doubtless worthy of a comment to highlight the role of the Donzère-Mondragon canal and its locks. In France, this is one of the canals that respects the size norms of Europe (11.40 m of draw, 90 m wide), which allows the transit and crossing of large barges all along the Rhine-Rhone. Its maximum flow rate of nearly 1900 m^3 s^{-1} allows both the André BLONDEL hydroelectric plant and the Tricastin nuclear power plant to be supplied (see figure 8.8). In the immediate vicinity of these works are the highest locks in Europe: 23 m high. The dam is also equipped with a fish ladder to allow the normal passage in both directions of the local fauna.

In this category of dams along waterways, we should not forget to mention natural dams, which often form in narrow valleys from the solid materials brought by the river: stones of all sizes, large blocks of rock, alluvium, and tree trunks combine to create these structures. While a dam forms, the water pressure on the upstream face plays a useful role by facilitating the packing of material and thereby giving the structure a certain cohesion. However, as the water level rises, a limit may be reached where the pressure on the upstream face can destroy the natural dam. Thus, instead of prolonging its role of natural spillway, the dam suddenly fails, creating a flood surge that descends on the valley below. In chapter 4, when summarizing the geological history of the Mediterranean Sea, we evoked a natural event of this type on the gigantic scale of the Strait of Gibraltar. In 1219, a natural dam on the Romanche River, downstream of Bourg d'Oisans in Isère, France, failed and resulted in the flood of Grenoble; an example that still scars the collective memory. The dam formed from the accumulation of solid debris carried by the Infernet torrent to the south and the Vaudaine torrent to the north, near the current location of the Véna bridge. Here, where the valley narrows significantly, the alluvial fans of these two torrents combined and retained various solid materials, reaching the altitude near that of Bourg d'Oisans. The natural dam thus formed held back the waters and the alluvium of the Romanche for several centuries. The small and remarkably horizontal plain between Bourg d'Oisans and Rochetaillée, not far from the hamlet of Les Sables, is the alluvial vestige of this lake period. In fact, during the Middle Ages and until the failure of the natural dam, the village at the upstream extremity of the natural lake, which is now Bourg d'Oisans, was named Saint-Laurent-du-Lac (Saint Laurence of the Lake).

4. General structure of a hydroelectric facility

4.1. High-hydraulic head in mountainous regions

The visible parts of a high-mountain facility obscure more important and complex organs. These latter require a schematic description to understand how the overall system works. Consider first figure 8.9, which is not specific to any particular site, but strives instead to show the respective roles and complementarities of the main parts. Start with the reservoir (1) and the dam itself (2), which is shown in cross section and situated at high altitude (for example, at 1790 for the Chevril Dam at Tignes). A water intake is located somewhere on the upstream wall of the dam or on the neighboring slopes (3) (large facilities often have several water intakes), placed neither too low to avoid taking in too much mud and solid deposits, nor too high, otherwise only a small fraction of the reservoir would be exploited.

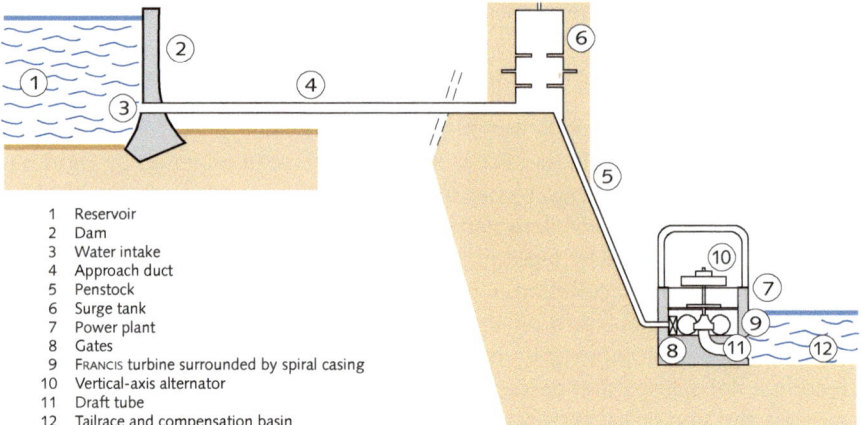

1	Reservoir
2	Dam
3	Water intake
4	Approach duct
5	Penstock
6	Surge tank
7	Power plant
8	Gates
9	Francis turbine surrounded by spiral casing
10	Vertical-axis alternator
11	Draft tube
12	Tailrace and compensation basin

Figure 8.9 - Schematic diagram showing functioning principles of a hydroelectric facility with a high-hydraulic head.

The water intake connects to long approach conduits (4) and that pierce through the mountains with mild slopes. These bring the water from the reservoir to the vicinity of the power plant. Geological studies done beforehand must demonstrate that these conduits, which are permanently filled with water, are robust and reliable. At Tignes, which is a particularly complex facility, several water intakes and conduits were constructed to carry water taken from the reservoir toward various power plants, the largest of which is at Malgovert, 15 km downstream. To limit the speed of the water in these conduits and thereby minimize the erosion of the walls, their diameters can be up to several meters. In these large-diameter conduits, water flows slowly and under moderate pressure, but the moving mass is so large that no gate can possibly stop this flow in the several minutes necessary to modify electricity production to meet demand. Thus, the conduit does not empty directly into the power plant and the turbines but into a penstock (5) made of steel that is solidly anchored in rock.

The flow is encased by the penstock, which is much shorter than the approach conduit and much smaller in diameter (normally several tens of centimeters). The penstock guides and confines the water through a vertical drop (called a *hydraulic head*) of several hundred meters to over 1000 meters at times, until it enters the turbines. The penstock is generally in the open air and more-or-less hidden by vegetation, depending on the altitude. In the facility at Tignes, the penstock at Malgovert must stretch 1500 m to reach the power plant (7), which is at an altitude of 750 m, or 1000 m below the reservoir. It is anchored to the rocky slopes at an average angle of 45° above the horizontal. With this difference in altitude, the internal pressure of stationary water at the base of the penstock can attain 10^5 hPa, or 100 times greater than atmospheric pressure.

Despite the necessity of rapidly closing the gate, the enormous mass of water contained in the approach conduit and in the penstock must gradually come to a halt. To accomplish this, a surge tank (6) is connected at the junction between the approach conduit and the penstock. The gates (8) are situated at the entrance to the power plant (7), immediately downstream of the penstock. When they close, the slow-moving water in the approach conduit fills the surge tank, which can expand sufficiently to stock the water as it arrives. The water level thus climbs above the equilibrium point until the pressure difference forces it back toward the reservoir. Oscillations ensue, which the surge tank is designed to progressively dampen with alternating narrow and wide sections. Reciprocally, when the gates open, water stocked in the approach conduit cannot instantly begin moving; the water from the surge tank thus steps in to provide the initial flow in the penstock toward the turbines. The depression thus created at the downstream end of the approach conduit then progressively accelerates the water. Again, oscillations ensue before the system reaches the steady state. The surge tank also drastically reduces a sort of shock on the gates that is linked to mechanisms of sound-wave propagation in the system; this is called a *water hammer.*[6] The surge tank is thus an important component that is conceived and designed by specialists who must

6   A water hammer is a shockwave generated by a sudden variation in the speed of a liquid flowing
     in a conduit, as can happen following the rapid opening or closing of a gate or after a pump is shut
     off. This violent shock is translated into significant vibrations and a deafening, characteristic noise
     that can be heard in facilities that are not equipped to counter this effect. It can damage the con-
     duit or the gates and, in extreme cases, lead to the rupture of the conduit. Water-hammer arrestors
     are designed to suppress this shockwave by increasing the time over which the water speed varies.

consider not only the amplitude and duration of these oscillations but also the demands of the site and of the materials used.

Finally, immediately after opening the gates (8), which are technical marvels themselves, the flow starts toward the turbine (9). In facilities in mountainous regions, the flow rate remains modest and the power is essentially due to the hydraulic head, which is on the order of 1000 m. In these conditions, PELTON turbines are used, as illustrated in figure 8.10. These consist of wheels mounted on a vertical or horizontal axis with shallow spoon-shaped buckets mounted on their periphery and that are impacted by water jets emanating from very specialized injector nozzles. Wheels with vertical axes generally use a larger number of injectors (up to six) than wheels with horizontal axes (at most two). The jets resemble somewhat those used by firemen to project water from their firehoses over long distances, but they are much more powerful and their flow rate can be tuned very precisely by using *spears*, which are visible in figure 8.10. Because they must withstand the pressure compiled by the large hydraulic drop of the penstock, their anchorage in the foundations of the power plant is studied very carefully. Each jet that strikes a blade exerts on it a constant force equal to the change in momentum of the water and, together, the jets exert a torque on the wheel, thereby releasing their energy to it. Under the wheel, the water drains into an open-air collector and carries on toward the river. Downstream from a FRANCIS turbine (see below), the water passes successively through a draft tube (11), which reduces its speed and the associated kinetic energy, and then through a tailrace (12), which is conceived to spread the variations in flow rate over rather long periods.

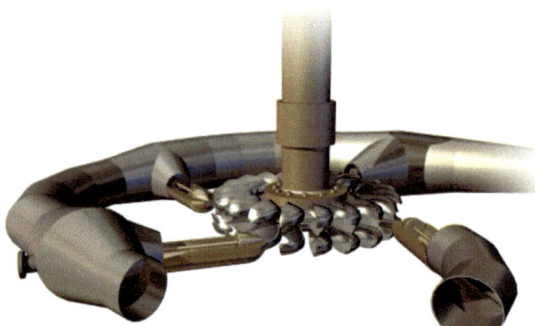

Figure 8.10 - Three-dimensional schematic drawing showing the principle of the PELTON turbine with a vertical axis designed for large hydraulic heads ranging from 200 to 1800 m. Note the wheel equipped with a series of buckets (20 in this case) served by four injectors equipped with spears (bronze color). [© 2017 General Electric Company]

These types of facilities first appeared at the end of the 19th century, when the Grenoble engineer and paper industrialist Aristide BERGES had the idea to capture water from some of the lakes in the nearby Belledonne mountains and bring it by penstock to turn the defibrators in the pulp mills at Lancey. The first penstock was built in 1869 and provided a 200 m hydraulic head and a modest flow rate of about 8 m^3 s^{-1} onto a water wheel (the ancestor of the PELTON turbine, which was invented 10 years later). This penstock, constructed from sheet iron as was written in the local daily, was 450 m long, 40 cm in diameter, and raised serious concerns and controversies that were echoed in the regional press: *such tremendous pressure in a single pipe!* writes the same daily paper. Its success thus resounded considerably and encouraged BERGES to undertake a second, more ambitious, project consisting of a penstock with a 480 m hydraulic head, which was completed in 1882.

In time, the penstock and turbine replaced the norias and animal work that were used up to that point to shred the spruce logs from the local forests and produce the cellulose fibers from which the pulp was fabricated. This turbine furnished so much energy that it was used first to light the village of Saint-Mury-Monteymond and then, as an experiment, Grenoble itself on July 14, 1882. It also provided energy to run a tramway from Grenoble to Chapareillan (42 km) in 1896, and finally to light at low cost the entire Grésivaudan Valley. During the World's Fair in 1889, in juxtaposition to the energy derived from coal mines by the German paper industry, Aristide BERGES coined the term *white coal* (a French equivalent of *hydropower*) to describe this type of energy, which comes from the glaciers and mountain torrents. We should also recall that the inhabitants of the Grésivaudan Valley, notably those living near the penstock, filed several lawsuits to try to stop the construction. These continued after the death of BERGES in 1904 until they were dismissed in 1906.

4.2. Medium hydraulic head

When the hydraulic head is of the same order of magnitude as the height of the dam, which is generally less than about 100 m, the facility can be much more compact, with the hydroelectric plant situated immediately downstream of the dam. The Génissiat Dam (see fig. 8.7) is typical of this type of dam. Note the absence of approach conduit, since the water intake in the upstream face of the dam directly supplies the penstock, which is also integrated into the dam. Compared with a large hydraulic head, much less liquid mass must be put in motion or stopped. This type of facility can thus do without a surge tank.

After passing through the gates, the water arrives at the turbine after passing first through a spiral casing designed to make the water rotate around the periphery before accessing the wheel. The turbine shown in figure 8.9 is a FRANCIS turbine (fig. 8.11),[7] with its vertical axis. The buckets and the massive disc below which

---

7   James Bichens FRANCIS was an English-born American engineer who, in the 1860s, popularized the turbine that now bears his name. The French engineer Jean Victor PONCELET originally conceived this type of turbine in the 1820s. The process was patented by the American engineer Samual B. BUCKENS.

they are attached must be very carefully designed both to optimize the energy conversion efficiency and to avoid cavitation,[8] which is the sworn enemy of turbines and pumps because it can damage them if they are ill conceived. We should add, however, that a well-designed turbine can last over 50 years. Downstream of the turbine in figure 8.9, we see the draft tube (11), which is another important part of the system. Its role is to strongly reduce the water speed (because the flow rate in the draft tube must be constant, increasing its cross-section results in decreasing the water speed) before it reaches the tailrace (12). Returning the flow to the river with a relatively low speed, the draft tube (11) decreases the loss in kinetic energy, which is proportional to the square of the speed of the water released into the tailrace, thus participating in the efficiency of the system.

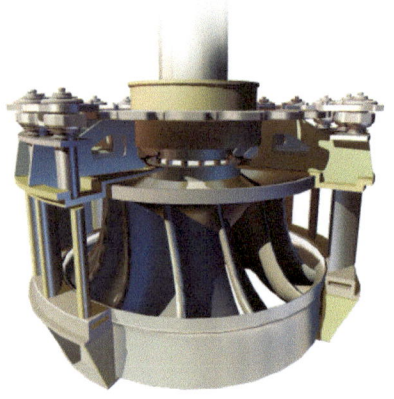

Figure 8.11 - Three-dimensional schematic drawing showing the principle of a Francis turbine with a vertical axis, designed for hydraulic heads ranging from 40 to 700 m. Around the wheel, note the series of directional blades that can be oriented to control the flow rate. [© 2017 General Electric Company]

4.3. Facilities with small hydraulic heads

All along large rivers, such as the Rhone whose overall management has already been described, regularly spaced facilities may be found, called *run-of-river* or *low-head dams*, where a large flow rate carries power through a small difference in elevation in a canal that runs alongside the river from the dam to the cascade itself. The imperatives of navigation often require locks to be built next to the cascade and to the power plant. The complementary flow rate continues its path in the natural river bed, allowing floodwaters to be dispersed and expunged. This maneuver is done by adjusting the very large sluice gates between the piles of the bridge-dam. The aerial photograph of the Andre Blondel power plant in figure 8.8 gives an idea of the dimensions of such a facility.

---

8 Cavitation refers to evaporation provoked by a strong decrease in pressure with no significant decrease in temperature. The buckets of a turbine or of a pump are systematically subjected to this risk because one of their faces experiences an overpressure and the other an underpressure. These buckets are designed to avoid cavitation, which can occur at the trailing edge.

The gently sloping lateral canal is analogous to the approach conduit of high- or medium-head power plants, but is open to the air. These canals generally stretch over 10 to 30 km (for example, the canal for the Donzère-Mondragon Dam at Bollène is 24 km). As for medium-head facilities, the presence of the free surface means that no surge tank is required. After transferring its energy to the turbines, the flow in the canal is reunited downstream with the complementary flow in the natural river bed. At Bollène, the Andre BLONDEL Dam, named after the architect of this facility, boasts six KAPLAN turbines,[9] each capable of producing 354 MW, for a total energy output of 2140 GWh per year. These KAPLAN turbines (fig. 8.12, left) are mounted on a vertical axis that also serves the alternators mounted higher up. This type of turbine is part of the family of helical turbines, whose specificity is that the orientation of their blades can be varied about the horizontal axis of each blade, allowing them to be precisely optimized regardless of flow rate. Note the difference between these propellers, which have only three or four high-precision blades, and that of a FRANCIS turbine, which often has a dozen buckets: the former obstruct the flow much less and thereby offer less resistance to the flow, which is precisely what is needed for large flows rates.

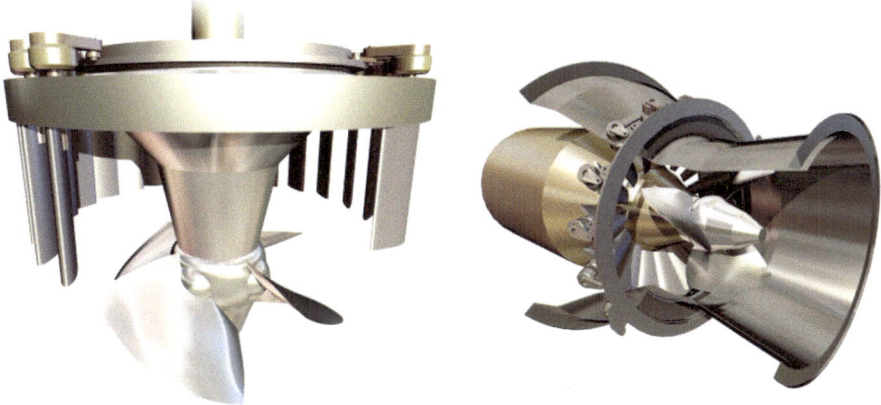

Figure 8.12 - Three-dimensional schematic drawing of turbines designed for small hydraulic heads. The left panel shows a KAPLAN turbine with a vertical axis, which is designed for heads less than 50 m; note in this case the wheel with three adjustable blades. The right panel shows a bulb turbine, also called a *horizontal KAPLAN turbine*; this is designed for heads less than 30 m. The warhead-shaped part at the upstream end contains the alternator. [© 2017 General Electric Company]

---

9   This type of helical turbine has adjustable blades and was invented by the Austrian engineer Viktor KAPLAN in 1912.

4.4. Other types of facilities

These three large families of hydroelectric plants briefly described above represent the vast majority of installations worldwide, although not all. We should also cite tidal-flow power plants, one of the most famous examples of which is the facility in the Bay of Rance in Normandy, France, close to Mont-Saint-Michel. This particular site was chosen because of its very large tidal ranges, often of the order of 12 to 14 m. The facility was inaugurated in 1967 and consists of a 750-m-long bridge-dam with a central portion of 332 m that is equipped with 24 helical turbines on horizontal axes, often called *bulb groups* (fig. 8.12, right). The machines have the peculiarity that they can operate in both directions, for an incoming tide or an outgoing tide, but their output is only viable for a quarter of a period, when the hydraulic head on one side or the other is about 4 m. The installed power is 240 MW and the energy produced is of the order of 500 GWh per year. After over 40 years of operation, this facility has revealed some nuisances, such as the ceaseless siltation of the bay and a modification of the seafloor. Other effects are observed that are inherent to withdrawing energy and were expected by the designers, such as a significant reduction in tidal range and a modification of some marine currents.

As for the small-scale facilities, we should cite microhydro power plants, which are constructed on streams with the necessary energy capacity, meaning essentially the combination of sufficient flow and hydraulic head to produce 100 kW of power. The water-use legislation of each country determines the conditions under which the water-rights holder[10] can construct such a facility and sell or use the power. A microhydro power plant is often a welcomed alternative to wind turbines or solar panels for supplying energy at isolated sites that are off the national grid.

5. Large port facilities

Among the works capable of containing large bodies of water, ports, with their breakwaters and piers, occupy a special place. They fulfill so many functions, which are, moreover, becoming ever more varied: harbor protected from storms, basin adapted to the expected tonnage of the ships, protection for the neighboring coastline, storage of merchandise, road and rail communication routes to serve the port, etc. In the time of the great explorers who left to discover the Indies, ports

---

10 *Water-rights* refers to the set of laws and regulations that determines the legal regime for waters; in other words, the conditions under which the holders of such rights can access the water and the measures they can take to protect the hydraulic resource. This definition reveals the objectives and the characteristics of water-rights that must consider both the quantitative aspects (management of the resource, regulation of its use) and the qualitative aspects (pollution prevention, treatment, penalties). In France, a law passed on January 3, 1992 sets out all the policies that aim to placate these diverse aspects.

were built in natural harbors, already well protected from storms. To these natural ramparts were added more protections made of wood, as well as a quay, also made of wood, which served to facilitate loading the ships. The dimensions of a large port were of the order of several hundred meters, enough to accommodate several schooners.

Economic development pushed both ship owners to conceive ever larger ships to reduce the cost per ton transported and port authorities to construct artificial harbors capable of receiving the large cargo ships that transport merchandise in voluminous containers. The quays have become immense container holds, equipped with powerful ship-to-shore cranes and complex systems of loading and unloading. One of the current records in this race to scale must be the Port of Los Angeles in California, USA (fig. 8.13). It covers an area of 39 km² along a 65 km stretch of the Californian coast. The annual tonnage transported increases steadily each year, depending on the commercial trade between the USA and the west; in other words, with China and Japan.

Figure 8.13 - Aerial view of part of the Port of
Los Angeles, showing its immense scale. [© USGS]

Several large construction projects were undertaken to give high-volume road and rail access to the various quays and warehouses where the merchandise is stocked. The most famous is no doubt Vincent THOMAS Bridge, which crosses high enough over the harbor water to allow the large container vessels to pass under its deck. It is over 1 km long with a main span of 457 m (fig. 8.14).

The quays and harbors of modern ports are constructed from concrete poured over a foundation of riprap, which solves the difficult problem of corrosion of concrete by salt water. The breakwaters that protect that from the incessant drumming of waves of all types and of all intensities are surrounded by riprap containing rocks sufficiently large to avoid being swept away. Natural riprap rock is also fairly frequently replaced by structures with four arms (tetrapods) made of reinforced concrete. These have the advantage of interlocking with each other while maintaining sufficient distance between themselves to accept the ebb and flow of the waves, whose energy they dissipate. Compared with the natural riprap, tetrapods also offer the advantage of being easy to remove and replace as soon as required by their erosion. In addition, modern ports also harbor very large dry docks and shipyards, where ships are built and corroded or damaged hulls are repaired.

Figure 8.14 - Vincent THOMAS Bridge, which spans over one of the access waterways for the Port of Los Angeles. The palm trees and cars, visible on the lower left, give an idea of the scale of this suspension bridge, whose main span is 457 m long and is sufficiently high to allow the largest container vessels to pass below its deck. [© Flickr.com/PRAYITNO Photograph]

Designing a new port along the coast of a developing country continues to raise a series of serious questions. How will this facility modify the marine currents? Having the new port fill with sand while the surrounding beaches empty of theirs would of course be unacceptable for all the stakeholders. Engineering consulting firms specialized in the design of these large waterworks now have modeling tools available that, with knowledge of the coast, the local seafloor, and the open-water currents, allow them to optimize the geometry of the basins, the quays, and the piers. Experiments with scale models, however, are still used to verify the numerical models because, even if the specialists are confident in their models and regularly implement them with success, their clients prefer to verify that the work meets their requirements before agreeing to the construction of the port.

Along this vein, the project currently under construction at Dubai in the United Arab Emirates is particularly original (see aerial view in fig. 8.15). The general goal is not to construct a port, but to create one or more islands linked by road and sea routes on which hotels and luxury residences can be built, the idea being to attract to this country of permanently blue skies a large fraction of the international luxury tourist trade. Beaches and pleasure ports are other attractive elements of the project. The marine currents around these artificial islands, as well as the eventual transport of sand into this part of the Persian Gulf, with its modified geography, were the subject of particularly intense study. Before construction began on this project, any eventual modification of the seafloor around the artificial islands was one of the questions studied with particular care and one that was closely followed during construction. It continues to be after construction, notably because of the huge scale of the project, which is so much larger than any previous port or coastal construction that no experience of similar scale exists to refer to.

Figure 8.15 - Peninsula in the form of a palm tree at Jumeirah, Dubai. The construction began in 2001, and completion was scheduled for 2013. The circular breakwater serves to protect the entire peninsula. [© NASA]

Conclusion

All these powerful dams and power plants that they supply, these irrigation canals and locks that allow navigation, these port and coastal facilities conceived for the benefit of the local economy—do they not seem foreign to the calm captured in song by Larmatine? And yet, just by their gigantic size, by their insertion into nature, by the control of floods and erosion of the banks, all these works educe a new poetry: that of the Hercules by Jean-Marie PIERRET (see fig. 8.4), but also that of the surroundings that they reflect in their waters.

At all times, as the demands of civilization grow and with the development of nations, man has needed to control his hydraulic resources, to tame the waterways and manage lakes and ponds. His initial goal was to satisfy the needs of fishing and agriculture and to protect himself from floods. The demands of industry that evolved into those of electricity in the 20th century led modern man to conceive large dams and hydroelectric plants that produce electrical energy that is then distributed over the power grid. In all their variety, in the mountains or along large rivers, the workable sites continue to attract attention for ever more integrated facilities that aim to satisfy complementary goals, such as producing energy, irrigating agricultural lands, navigation, road transport, or tourism. No sooner than one is completed is its impact assessed; a process in which citizens can participate through their representatives. This is done according to the laws and regulations of the state in question, or even in an international framework for waterways whose watersheds extend into several countries. And, with the examples of the flooded village of Tignes or with the Three Gorges Dam in China, we have seen that these debates can often finish in court, or even in front of the national parliament. Yet despite taking all these precautions, the risks of catastrophe remain present, and maybe even ineluctable.

Epilogue

One sees qualities at a distance and defects at close range.

(Victor HUGO, *Heap of Stones*)

R. Moreau, *Air and Water*,
DOI 10.1007/978-3-319-65215-3

T his stroll through air and water has to this point been done in a spirit of amazement in front of this two-fluid medium, which is immense, powerful, and complex and which sustains life on Earth. Still, before completing our trip, we cannot let pass in silence all the worries regarding its evolution that have been so forcefully expressed since the end of the 20th century.

1. What worries,
and on what are they based?

What, then, are the threats to life on Earth, considering how humanity uses these precious fluids that are air and water? Without question, the consequences of pollution and the lack of freshwater must be at the head of any inventory. Their analysis is the subject of detailed studies that nourish debates because the dangers to which they can expose humanity and all living species constitute serious political and economic issues. That said, the level of uncertainty that characterizes these threats remains relatively large because of the inherent difficulty of measuring these effects and the lack of knowledge of their effective impact. The extent of this uncertainty is such that it is often difficult to verify certain numbers, even those that are presented as data. As for the possible evolutions, the experts dress their predictions with significant escape clauses, and their conclusions can sometimes diverge. We therefore limit this brief discussion to a sort of summary of the state of affairs. The interested reader is invited to consult the specialized texts to better evaluate the risks, their eventual impact, and the stakes involved for society.

The ambiguity in fact begins with the definition of the word *pollution*. Because air is a gaseous mixture that has always contained particles in suspension, and because the oceans constitute the vast reservoir of nature (see Jules Verne citation in the beginning of chapter 5), where does pollution start? A certain amount of natural pollution has always existed,[1] fed both by uncommon events such as large volcanic eruptions, and by more discrete and quasi-permanent sources such as the dispersion of pollen. The question is thus more that of the anthropic evolution of this pollution, rather than of its actual state. Let us thus consider this double aspect and limit ourselves to extracting estimates on which the experts agree. We shall begin with air pollution, and then follow up with water pollution.

---

1   In the Middle Ages, air and water already contained various contaminants: our ancestors feared that these fluids were corrupted and attributed their epidemics and microbial infections to this pollution.

Concerning the shortage of freshwater, the uncertainty in the estimates of the current state of this resource and of the evolution of humanity's needs seem less large. The recent data are in fact the largely agreed upon and justify worldwide concerns, which we shall attempt to identify.

1.1. Air pollution

For over 10 million years, the ice of Antarctica has stocked and recorded all atmospheric pollution, with both its concentrations and its dates! It constitutes a physical memory; we find not only the contaminants emitted by volcanic eruptions but also the dust emitted by the mines and foundries of the Roman Empire. The analysis of this ice cap shows that a massive growth in contamination begins with the industrial revolution, which began in the middle of the 18th century, when metallurgy started to use coal instead of charcoal. In the western countries, this industrial evolution led to the well-known smogs such as that of London, which disappeared only at the end of the 20th century with the departure of the majority of the extremely polluting heavy industries. In China, on the contrary, the second half of the 20th century has seen industrial development induce the current pollution, which increases dangerously and without relent. The energy demands of this huge country are such that, almost every week, a new coal-fired power plant is commissioned.

The first concentrations to discuss are those of the greenhouse gases, which are largely responsible for global warming (climate change was already discussed briefly in chapter 1). Apart from water vapor, a natural component of air that is indispensable to all forms of life on Earth and is by far the largest contributor to the greenhouse effect, the most abundant greenhouse gas is carbon dioxide (CO_2). This gas is the end product of the combustion of all carbonaceous material—solid, liquid, or gas. Its concentration by volume is of the order of 350 parts per million (ppmv) and increases by 1.3 ppmv each year. Approximately 30% of the total greenhouse effect is attributed to this gas, compared with 60% for water vapor. The concentration of methane (CH_4), which comes from ruminants, flooded crops such as rice, and landfills, is of the order of 1.8 ppmv and increases by 0.0012 ppmv each year; its contribution to the greenhouse effect is about 5%. Ozone (O_3) in the troposphere[2] is said to contribute roughly the same as methane to the greenhouse effect, but the lifetime of this unstable molecule is less than a few weeks, whereas that of methane is of the order of 10 years and that of carbon dioxide is about 100 years. Despite its instability, ozone, which is produced by the decomposition of pollution from automobiles, is continually renewed, so that its concentration in the

---

2 The ozone in the troposphere is the result of photochemical reactions that decompose certain pollutants emitted by automobiles, notably nitrous oxides. Because of its short lifetime, ozone remains in the vicinity of where it was produced and essentially does not interact with the ozone of the stratosphere (see chap. 1) that, far from being a pollutant, protects us from ultraviolet radiation from the Sun.

troposphere is permanent and growing but restricted to the vicinity of the production sites; namely, the large urban areas and major road junctions.

We can also cite the various nitrous oxides, normally written as NO_x and which are produced by burning liquid or gas combustibles as well as by burning biomass (firewood in particular), carbon monoxide, which results from incomplete combustion, and sulfur dioxide (SO_2) and the various aromatic hydrocarbons, which are emitted notably by burning wood and organic matter in the open air. Finally, chlorofluorocarbons (CFCs), also called *freons*, were much used in the 20th century in refrigeration and air conditioning equipment because of their high heat-transfer capacities and escaped into the atmosphere after accidental leaks or the destruction of old appliances. In western countries their use was henceforth limited by regulations, which stipulated that they be replaced with hydrochlorofluorocarbons (HCFCs). Since the beginning of the 21st century, these regulations reduced the consumption of CFCs by 75% in western countries.

Air also contains numerous solid particles, or aerosols, which range in size from microns to several tens of microns, and which remain in suspension for long times (weeks or even months) precisely because of their small size. They often come from the carbonaceous soot emitted by incomplete combustion in diesel motors and in all fires involving wood, sawdust, or organic wastes. Finally, they precipitate back to the ground only when rain drains them from the sky (as we saw in chapter 3, these aerosols serve as seeds for forming raindrops and snowflakes). But those that fall on the ground and are dried by evaporation again form microscopic dust that turbulent winds can claim and push back up into the atmosphere. The only real way to extract these particles is via the oceans where they are first fixed into the surface biofilm before a fraction of them are very slowly transported to the sediments on the seafloor, with the remaining again being released into the air by evaporating sea spray. The consensus seems to agree that the danger of these microparticles comes from the fact that they penetrate into the lungs where they can provoke respiratory ailments, allergies, or serious lesions.

Finally, industrial activity has created new types of pollution by injecting into the atmosphere various contaminants that were not present before. First, without giving their small and relatively uncertain concentrations, are the biological pollutants emitted by the factory farms. We should also mention the metallic pollutants that differentiate themselves from the former by not being biodegradable; among them, the most toxic seem to be the heavy metals, notably mercury, lead, and cadmium. Curiously, we could note that the attempt to reduce pollution by automobiles, notably the production of NO_x, by requiring the use of catalytic converters has generated a new metallic pollution. This is linked to the dissemination in the atmosphere of platinoids (platinum, rhodium, and palladium), which are used as catalysts in chemical reactions that reduce nitrous oxides by transforming them into CO_2. Moreover, let us not forget the radioactive wastes that come from the open-air nuclear tests or from the accidents that have occurred in nuclear power plants.

For example, the cesium-137 that was released into the atmosphere during the Chernobyl disaster in 1986 is in the process of being reabsorbed, because its half-life is on the order of 30 years.

1.2. Marine pollution

We have seen in a preceding paragraph that one of the origins of marine pollution is the air, which deposits a biofilm on the surface of the seas. This is nothing but one of the multiple aspects of the exchanges between the atmosphere and the oceans, profoundly linked to those described in chapters 5 and 6, such as the generation of the thermohaline circulation by the trade winds, the formation of waves, and the creation of sea spray. The physicochemical nature of marine pollution is thus partly identical to atmospheric pollution, which includes all substances released into the air from the ground or oceans, as well as by volcanic eruptions or by plants or human activities.

The input of the rivers and streams adds another contribution, which comes from both the runoff from cultivated lands, which is loaded with fertilizer and pesticides, and the discharge from waste-water treatment plants for urban, agricultural, or industrial centers. The particular case of the nutrients lost by intensive agriculture, especially the nitrates and phosphates, merits a special mention because they lead to green algae and micro-algae. The former release sulfur hydroxide (SH_2), a gas that becomes toxic in the stomachs of animals that ingest it. The latter secrete toxins that can also cause death within certain marine organisms.

In addition, the motorized vessels that crisscross the seas contribute a very significant fraction of the pollution of the marine air. In 2004, over 90000 vessels and more than 101 gross registered tons were in circulation, which consumed nearly 217 million tons of fuel, essentially heavy fuel oil. Unlike the fuels destined for land or air vehicles, marine fuel is not desulfurized. This maritime traffic is thus responsible for a significant pollution via the acidification of the emissions of sulfur dioxide (SO_2), which are progressively transformed into sulfuric acid (SO_4H_2), of which 16 million tons was added to the atmosphere in 2004.

Finally, all the maritime accidents, losses of hydrocarbons or other cargo, ejection of diverse wastes, and sunken ships, constitute another source of pollution. The deposits on the seafloors can sometimes infuse into the loose sediment and thus be neutralized, but significant amounts of waste remain suspended and are transported by the marine currents over immense distances. The most cited example is the Pacific Trash Vortex,[3] which contains between 30000 and one million items per square kilometer and spreads, at a shallow depth, over 20 million km^2.

---

3 The American oceanographer Charles MOORE recently measured the concentration of debris in the Pacific Trash Vortex within the first few meters from the surface, thereby providing relatively reliable numbers.

This ensemble of marine pollution proves difficult to measure and thus to control. To illustrate this uncertainty, note that one of the markers used is the toxicity of shellfish that retain toxins by filtering seawater. As nonquantitative as it may be, this indicator is what revealed the rapid increase in marine pollution seen during the 1990s, no doubt due to the discharge of ballast water[4] by vessels that travel the globe and spread toxic algae. This pollution was considerably reduced in the first decade of the new century, after regulations obliged crews to maintain a record of their ballast water.

1.3. Freshwater resources

To understand how sensitive this issue is, consider some essential figures. The volume of inland water is of the order of 35 million km^3, which pales in comparison to the 1350 million km^3 in the oceans. And, among these 35 million km^3, only nine is available for human use (i.e., whose salinity is less than 3 g kg^{-1}); about one quarter of the total. This water is essentially contained in the water tables. Moreover, the planet now hosts and must feed 7 billion inhabitants who consume 42 000 billion m^3 per year, or about 6000 m^3 per year per inhabitant. The growth of the population is expected to lead to 9 billion individuals by 2030. With the same 42 000 billion m^3 per year, the consumption must decrease to 4650 m^3 per year per inhabitant, which represents a decrease of one-third. To properly appreciate these numbers, one must also consider that irrigation currently accounts for about 70% of this consumption and should rise by 15% to 20% for agriculture to be able to feed humanity in the upcoming decades.

These numbers give an averaged view of reality, but considerable inequalities exist between different regions of the world, to the point of sparking veritable catastrophes, such as that ongoing in the Horn of Africa (Somalia, Ethiopia, Djibouti, and Kenya) since the beginning of the 21st century, following several decades of uninterrupted drought. According to the United Nations Children's Emergency Fund (UNICEF), which is the main humanitarian agency of the United Nations (UN) and whose mandate is to help children, ten million people are currently subjected to a serious water shortage and suffer from malnutrition. The UN places the alert threshold for freshwater resources at 1700 m^3 per person per year. In some desert regions, people apparently already live under this threshold, and desertification, which concerns today 40% of the land area, should intensify because of climate warming. Henceforth, conflicts between states separated by river borders seem to boil beneath the surface. According to the UN, freshwater could become a commodity more precious than oil, which has led some observers to call it *blue gold*.

---

4   Ballast water is used to balance ships, notably modern double-hulled oil tankers, after a partial offloading. The replacement of waste water by clean water is now regulated by the International Maritime Organization.

2. To conclude our journey

Upon receiving sunlight, the fluid terrestrial media of air and water become restless and start moving to redistribute the energy over the entire planet according to their own rhythms in which we can detect almost all possible timescales, from much less than milliseconds to several millennia. Their dynamics, thermal properties, changes of state, and the interactions between the atmosphere and the oceans all result from subtle mechanisms, some of which continue to be the source of serious questions for the scientists that devote themselves to understanding and modeling them. A few central ideas have been extracted to attempt to gather them together and understand their main points.

This two-fluid medium is precious because it preserves the life forms that we know and allows them to evolve.[5] During the centuries, we humans have managed to know it and understand it better and better, and often to domesticate it to limit its dangers and its most extreme hazards. We simultaneously use it, pollute it, and protect it. We are thus responsible, which brings us to questions ourselves about its future, notably on the scale of the 21st century, which appears to be the scale on which the actions of the present generations can act. However, our investigation of air, water, and their interactions has revealed fearsome challenges with significant uncertainties that council the greatest prudence to all those who would make predictions.

After the atrocious cataclysms that humanity unleashed on itself during the 20th century, we live in a surprising period, marked both by a renewed interest in nature and by the expression of serious concerns for her future. After believing that science and technology could contribute to the well-being of humanity, we find that, at least in western countries, some of us are beginning to doubt it. Over the last few decades, many of the brightest spirits have turned away from science, attracted by other solicitations, at the end of high school or after their university studies. And yet, because our own worries take the form of scientific questions, how will humanity answer them without investing heavily in science?

At all times, each great advance along the path of knowledge has caused thinkers to question and fear a possible harmful use of the new knowledge. The conquest of *terra incognita* is in fact an adventure; as such, it is not exempt from danger. Nevertheless, since the appearance of *Homo sapiens*, humanity has crossed the worst catastrophes, not without suffering the consequences, but each time being able to rebound and start a new stage in its development. No one knows what the distant future holds for humanity, but it is obvious to all that ignorance would be the worst of all attitudes.

---

5   On Earth, the life expectancy of a species is estimated to be between 10 and 100 million years. According to specialists in biodiversity, only one out of a thousand species that has ever existed is still in existence, and the Holocene Era in which we live is one of the five periods of the most massive extinctions ever recorded.

To conclude, we propose to the reader two thoughts to assist in your own thinking. The first is borrowed from the philosopher Gaston BACHELARD: *Man wants to see. Curiosity energizes the human spirit.* It allows us to hope that humanity will never cease its quest for new knowledge. The second, which highlights the novelty of the challenges confronting today's generations, is extracted from a recent book by Robert DAUTRAY and Jacques LESOURNE:[6] *The 21th century is not like the others.*

---

6 *L'humanité face au changement climatique* by Robert DAUTRAY and Jacques LESOURNE, Odile Jacob Sciences, 2009, p. 134.

Appendix

Instabilities and turbulence

You never step twice into the same river.

(HERACLITUS)

© Springer International Publishing AG 2017
R. Moreau, *Air and Water*,
DOI 10.1007/978-3-319-65215-3

A s HERACLITUS testifies, our ancestors have known since antiquity that natural flows are essentially nonreproducible. In the atmosphere, in the oceans, and in all waterways, diverse types of hydrodynamic instabilities are at work that explain the random characteristics described in the preceding chapters. Certain instabilities linked to variations in density result in the sudden appearance of movement in a fluid initially at rest. Others, linked to the existence of shear interfaces between flows with different speeds, allow us to understand the appearance of waves along these interfaces. Moreover, the unstable evolution most often encountered is no doubt that which leads via complex paths from the laminar regime of the flow of a homogeneous fluid to the fully developed turbulent regime. The fluctuations of these turbulent flows cannot be reduced to a detail that we can neglect in a first approximation; on the contrary, they are significant enough to strongly modify the properties of the fluid. In particular, their capacity to dissipate energy, to decelerate movements, and to mix the fluid is systematically found to be superior to that of the laminar regimes, and sometimes by several orders of magnitude.

In this appendix, several keys are offered so that the reader may not only recognize these instabilities and this turbulence in atmospheric or marine flows, but may also perceive the physical mechanisms behind them and, in certain cases, to even sense their properties and consequences. In line with our original choice, we exclude from this summary the theoretical developments that allow us to describe with precision these characteristics. The content of this appendix is thus limited to presenting these mechanisms, evaluating their respective parameters, justifying the transition thresholds between stable and unstable regimes, and giving several orders of magnitude that are useful for observing them.

We will find that describing these instabilities and any subsequent turbulence that develops does not require us to specify the fluid in question, be it air, water, or other, because these are universal characteristics. This is the main reason we have chosen to present them in this appendix, which may be viewed as supporting material for the eight preceding chapters.

1. The sudden appearance of movement

1.1. RAYLEIGH-BÉNARD instability

Imagine a fluid at rest and let us see what mechanisms can explain why, suddenly, it begins to move. The prototypical transition of this type is the convective instability of a fluid layer heated from underneath, which is known as the RAYLEIGH–BÉNARD instability [1], illustrated in figure A.1. This instability appears when the lower part of such a layer is sufficiently warmed with respect to the upper part. If, for whatever reason (and such reasons always exist), certain columns of fluid become warmer than others, even imperceptibly so, then this perturbation of the initial stationary state may be amplified to the point that it attains a quasi-stationary amplitude. The fluid then goes through a transition marked by zero initial motion transitioning toward a new regime in motion, which we will attempt to understand.

When a fluid layer is heated from below or cooled from above, columns appear that are slightly warmer than the ambient fluid and others appear that are slightly cooler. No sooner does this happen than the fluid delimited by these two columns starts to move spontaneously, like a sort of roller. Several such columns and rollers are illustrated in figure A.1. The real weight of the warm column, which is dilated and thus lighter than the ambient fluid, is less than that of the cold column, which in contrast is heavier. Although their average weight is balanced by the forces of hydrostatic pressure, these differences are not and constitute additional vertical forces directed alternately upward and downward.

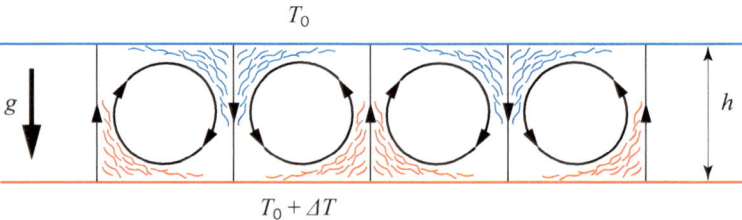

T_0

g h

$T_0 + \Delta T$

Figure A.1 - Schematic illustration of the mechanism behind the RAYLEIGH–BÉNARD instability in a fluid layer situated between a warm boundary (lower boundary with temperature $T_0 + \Delta T$) and a cold boundary (upper boundary with temperature T_0). The columns that drain the warm fluid from the lower boundary are lighter than the columns that drain the cold fluid from the upper boundary. Each domain between a pair of alternating warm (red) and cold (blue) columns is thus subjected to a driving torque.

On the scale of the cells shown in figure A.1, these forces form a driving torque τ_d, which results in the motion of the roll bounded by the two columns. The counteracting mechanism is friction, which acts at the boundaries of this rotating fluid domain and which adds a torque τ_f of opposite sign with respect to the driving torque. Of course, friction is linked to the viscosity ν of the fluid. The instability appears when τ_d becomes greater than τ_f. This classic critical condition is often

---

1   Henri BÉNARD (1874–1939) was a French physicist and professor at the University of Paris. He is credited with the discovery of the experimental analysis of convective structures that bears his name: BÉNARD cells. John William Strutt RAYLEIGH (1842–1919) was an English physicist and professor at the University of Cambridge. He was interested in all the physics of his era and notably laid the foundations of the theory of this convective instability. He was awarded the NOBEL Prize in Physics in 1904.

expressed in the following form: the instability develops if the RAYLEIGH number,[2] which is proportional to the ratio of the torques τ_d and τ_f, exceeds a critical threshold that can be determined with precision and that is typically of the order of 10^3. This threshold depends on the nature of the upper and lower boundaries (smooth or rough wall, free surface, or a combination of both), which determines the value of the torque τ_f; for example, between two smooth walls, the threshold is 1708.

Note thus that the condition that must be satisfied to trigger the instability demands that the temperature difference between the two limiting planes exceeds a critical value that depends on the thickness h of the fluid layer, on gravity, and on the properties of the fluid. This remark begs the following question: how can we estimate the temperature difference between the alternating warm and cold columns that drive this vortex? At most, they are of the same order of magnitude as the temperature difference ΔT imposed by the limiting planes, because the warm fluid near the lower plane is swept up toward the top of the warm column, whereas the cold fluid near the upper plane is drained by the cold column. This estimate is well suited for fluids that are poor thermal conductors, but it is not well suited for those that, on the contrary, are good thermal conductors, because the difference in temperature between the columns is reduced by thermal conduction across the rolls between the columns. Let us simply retain that the temperature difference between these columns decreases with increasing thermal conductivity of the fluid. This explains the presence of the thermal diffusivity α[3] in the denominator of the RAYLEIGH number: the larger is this diffusivity, the smaller is the temperature difference between the warm and cold columns, the ratio of the torques τ_d and τ_f, and the RAYLEIGH number itself.

In a 1-m-thick layer of air, the minimum order-of-magnitude temperature difference to attain the critical threshold and trigger the instability is about 10^{-5} °C; in other words, almost zero. This comes from the fact that this gas expands easily, is not very viscous, and is a poor thermal conductor. Air is thus very easy set in motion with this mechanism, even for very small temperature differences, provided the heat comes from below.

---

2   The RAYLEIGH number is expressed as $Ra = (g\beta\Delta T h^3)/(v\alpha)$, where g is the acceleration due to gravity, β is the coefficient of thermal expansion, v is the kinematic viscosity, and α is the thermal diffusivity. The other quantities are defined in figure A.1. When both walls are solid and smooth, its critical value at which the instability appears is $Ra_{crit} = 1708$. When one or the other of the two limiting planes is a free surface, it is smaller, but remains of the order of 10^3.

3   The thermal diffusivity is the ratio $\alpha = \kappa/\rho C_p$, where κ is the thermal conductivity (κ is about 0.1 W m^{-1} K^{-1} in air and 0.6 W m^{-1} K^{-1} in water), ρ is the density (ρ is about 1.2 kg m^{-3} for air and 10^3 kg m^{-2} for water), and C_p is the specific heat capacity of the fluid at constant pressure (C_p is about 10^3 J kg^{-1} K^{-1} for air and 4.18×10^3 J kg^{-1} K^{-1} for water). As a result, the thermal diffusivities of air and of water are about 0.8×10^{-4} m^2 s^{-1} and 0.14×10^{-6} m^2 s^{-1}, respectively. Note their purely kinetic character, like that of the kinetic viscosity, which leads to the PECLET number ($Pe = VL/\alpha$), which is exactly analogous to the REYNOLDS number ($Re = VL/v$). The PRANDTL number ($Pr = v/\alpha$) characterizes the ratio between the capacity v of the fluid to diffuse momentum and its capacity α to diffuse heat.

The consequence of this convective instability is the considerable growth in heat flux exchanged between warm and cold media situated on opposite sides of a fluid layer. Multiple examples of this are found in our apartments or houses and in the outdoors, even if the boundaries of the fluid domains generally have much more complex geometries than the horizontal planes shown in figure A.1. Thus, around a household heat source, the upward movement of air is normally significant enough to evenly distribute the heat throughout the room. Another example that each of us can observe is the transition from the quite calm of the morning air cooled from the bottom to the agitation that appears as soon as the first rays of sunlight start to heat the ground which, in turn transfers this heat by conduction to the lower layers of the air.

The preceding estimate indicates that, in the atmosphere, the RAYLEIGH number can attain values much greater than the critical value—several orders of magnitude greater. Thus, the atmospheric layer warmed by the ground is far from the critical threshold, so that the convective motion is not organized in rolls as in figure A.1 but becomes turbulent. This allows an efficient transfer of the Sun's energy while minimizing the temperature difference between the lower and upper layers. In general, this turbulent movement depends strongly on the dimensions of the region concerned, which enters into the expression for the RAYLEIGH number through the factor h^3.

Let us return now to the case of a horizontal fluid layer between two plane parallel walls. Slightly above the critical threshold, if the region concerned has a relatively large horizontal dimension, the convective movement is organized into cells spatially arranged in a well-defined periodic manner. The form of the cells depends on external parameters; parallels rolls, hexagonal cells, or other structures altogether may be observed. On the contrary, if the region of interest is small, as can happen in laboratory experiments, a general circulation can develop that occupies the entire container, either with a single ascending central current that descends at the periphery, or with an ascending column on one side of the container and a descending column on the other. Farther from the threshold, the movement may be manifested by warm, rising, and fairly localized plumes or cold descending plumes.

When the boundaries of the region of interest have a complicated morphology that differs significantly from a horizontal plane, motion is always present in the nearby air unless the conditions are isothermal. This is due to the fact that, along a portion of a warm, inclined wall, the nearby fluid is lighter than the fluid farther away but at the same height. This nearby fluid is thus systematically carried upward by the buoyancy force, which cannot be balanced by the forces of pressure. This is the case for the ground, which is never perfectly flat and is very deformed by its cover of vegetation that is cooled by evaporation. It is also the case above a radiator where turbulent motion can be detected by eye in the air thanks to the variation in the index of refraction, which is linked to the temperature. This weak turbulence thus suffices to perturb the normal transparency of isothermal air.

In the salt water of the oceans, the density depends on more than just the temperature; it also varies with salinity. Moreover, near about 3.98 °C, it does not vary monotonically with temperature (see chap. 5). The result is that the convective perturbations can develop a much more complex organization than those discussed heretofore, notably by forming several sublayers placed one above the other. The study of these situations becomes particularly complex because the positions of the interfaces between the sublayers are themselves unknown. We simply note that the surface water of lakes and in the oceans, which is warmed from above by sunlight, generally is not subject to the RAYLEIGH–BÉNARD instability. We shall see below what mechanisms generate the waves that constantly animate these surface waters.

1.2. RAYLEIGH-TAYLOR instability

Imagine now that the domain at rest consists of two homogeneous fluids of different densities and situated one above the other on opposite sides of a planar horizontal interface. If the heavier fluid is below the lighter fluid, all perturbations at the interface are opposed by the weight, which brings the system back to the initial state. However, if the heavier fluid is above the lighter fluid, any eventual perturbations tend to be amplified by weight (fig. A.2); the heavier fluid tends to pass beneath the lighter fluid by forming somewhere a descending finger, whereas elsewhere a finger of the lighter fluid rises so as to keep the total mass constant. The only counteracting mechanism is the surface tension (see insert I3.2), which opposes the destabilizing force of weight and can prevent perturbations from amplifying and thereby inhibit the development of the perturbation. This is the same as saying that only capillarity can stock the energy supplied to the system by the force of weight by increasing the interfacial area. The consequence is that this arrangement stabilized by surface tension can only be found at very small scales.

Quantitatively comparing the stabilizing and destabilizing effects is actually quite simple. Because the surface tension is proportional to the curvature of the interface, the most unstable perturbations are those that minimize this curvature; in other words, those with maximum length. These are the perturbations that stretch over the entire length L of the container holding the two fluids. Let us evaluate the pressure difference between the points M and M' in figure A.2. The pressure difference due to weight must be proportional to the product $gh\Delta\rho$, where g is the acceleration due to gravity, h is the amplitude of the perturbation, and $\Delta\rho$ is the density difference between the two fluids. The pressure difference due to surface tension is proportional to the ratio σ/R, where σ is the surface tension and R is the radius of the curvature of the perturbed interface. Note that geometry obliges this radius to be of the order of L^2/h. Finally, for the pressure at point M' to push this portion of the interface downward against the resistance given by the surface tension requires that $g\Delta\rho$ be greater than σ/L^2; note that this condition is independent of h. This instability condition means that the Bond number[4] must exceed a critical threshold of the order of 40. Above this threshold, during the development of this Rayleigh–Taylor instability, we can see a long finger of heavy fluid descending toward the lower part of the container, while another finger of lighter fluid rises toward the upper part.

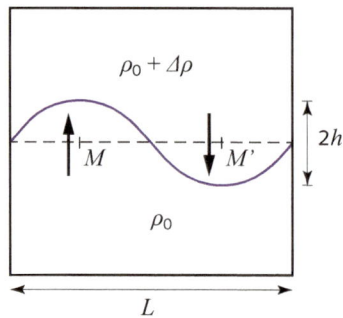

Figure A.2 - Rayleigh–Taylor instability at the interface between two fluids. The density of the upper fluid is greater by $\Delta\rho$ than that of the lower fluid.

In the particular case of a water-air interface (i.e., water on top, air on the bottom), the densities are so different that perturbations having a typical length of the order of centimeters or more are always unstable. They thus develop almost systematically. However, on very small scales of the order of a millimeter or less, such as that of an eye dropper which we know can maintain a hanging drop, the user must clearly apply an overpressure with his fingers to force the drop to grow and detach.

In the case of water-water interfaces, such as that of the thermocline in the oceans (see chap. 5), the difference in density is much less, by several orders of magnitude. If we accept that the concept of surface tension remains valid, which is debatable because the thermocline has a finite thickness that is much greater than the molecular scale, we arrive at critical lengths on the order of tens of meters. Despite its debatable validity, this approximation does have a certain significance: the formation of fingers could only happen on rather large scales, of the same order of magnitude as the thickness of the fluid layer above the thermocline. However, except in winter when the temperature varies little in the vertical direction, the deep water is colder and heavier than the water above the thermocline, making it almost impossible for a Rayleigh–Taylor instability to appear.

In chapter 3, we discussed the beginning of a tornado in the form of a finger that descends below a mass of clouds. In so doing, we discussed the interface between the bottom of a dark cloud and the transparent air below it. Above this interface, the fluid is a nonhomogeneous mixture of suspended water drops and, below the interface, the fluid is air saturated with water vapor. The fact that the interface is clearly visible over long distances implies that its equilibrium is possible; in other words, the average density of the cloud near the interface is less than that of the

---

4   The Bond number is the ratio $gL^2\Delta\rho/\sigma$. Its value serves to compare the respective influence of gravity and surface tension.

air just below it, allowing the air to support the cloud. To explain the formation of the very localized descending finger, we considered the rotation near the axis of the central vortex that formed in the cloud. This rotation generates an additional centripetal depression that strongly increases the local condensation, leading to the formation of large liquid drops that are heavier than those situated in the surrounding area. When the air, which supports the main part of the cloud, can no longer carry this central heavier part, it descends and forms a rotating finger that creates the tornado structure as soon as it touches the ground.

2. Sheared interfaces: the KELVIN-HELMHOLTZ instability

Another instability often seen in the atmosphere and oceans and known as the KELVIN–HELMHOLTZ instability is generated by shear between two fluid masses that circulate at different velocities. This instability is represented in figure A.3 in a reference frame moving at the average velocity of the two layers. The figure assumes that the fluid above the interface flows to the right, whereas the fluid below flows to the left. Imagine a sinusoidal perturbation at the interface. Above point N is a protrusion that forms a small obstacle, forcing the lines of flow to tighten, which means that the velocity increases locally and the pressure decreases. Conversely, under the concave part of this protrusion (i.e., below point N), the lines of flow have more space and thus spread out, meaning that the velocity decreases locally and the pressure increases. The situation is exactly the opposite near point Q. Thus, any sinusoidal deformation of the interface systematically leads to a difference in pressure whose sign is such that the deformation is amplified.

It is interesting to interpret the KELVIN–HELMHOLTZ instability in terms of a vortex, or vorticity,[5] because this leads to a way to recognize it in the atmosphere. The rapid variation of the relative speed from $-V$ to $+V$ in traversing the interface is in reality spread over a thin boundary layer of thickness δ. Without the instability, the interface is characterized by a shear, or by a vorticity, whose order of magnitude is V/δ. This vorticity is transported by the moving fluid. In the region of the hump MNP, on the side of the faster-moving fluid, the vortex is shifted to the right toward point P because the perturbation in the flow at N is directed toward P. In the region of the trough PQR, on the side of the slower-moving fluid, the vortex is shifted to the left, again toward point P, because here the perturbation in flow is oriented from Q toward P. Thus, the vortex that is initially spread over the entire shear interface becomes progressively concentrated near the inflection points such as P, and away from the other inflection points such as M and R.

---

5   The vortex is defined as half of the curl of the velocity: $\boldsymbol{\omega} = \boldsymbol{\nabla} \times \mathbf{V}/2$, which is a vector. The coefficient ½ ensures that the magnitude is exactly equal to the local angular speed of a fluid particle. The magnitude $\omega^2/2$ is called the *vortex intensity* or the *vorticity*.

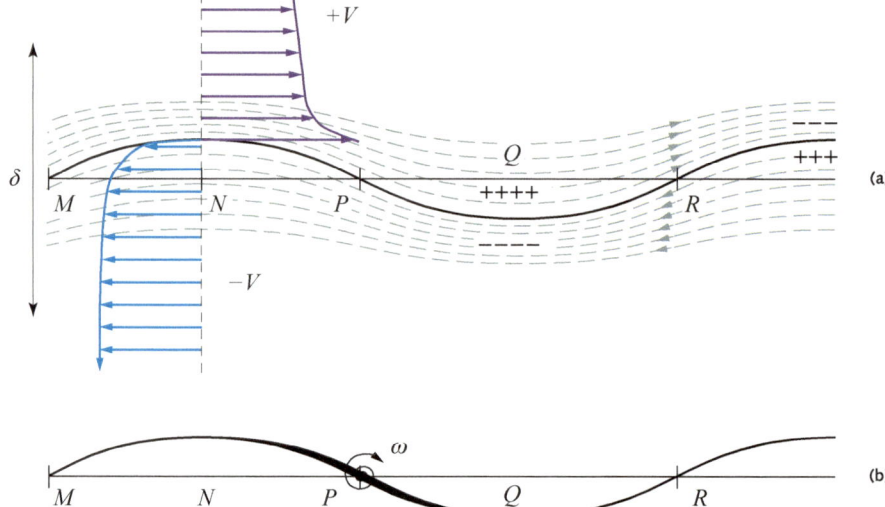

Figure A.3 - Illustration of the KELVIN–HELMHOLTZ instability between two fluid layers with shear. The flow is represented in a reference frame moving to the right at the average speed of the two flows. The fluid above (below) the interface is moving to the right (left). **(a)** Schematic illustration showing the pressure difference that is accentuated between the humps and the troughs (+++ indicates an overpressure, − − − indicates an underpressure). The dashed lines indicate the streamlines (tangent to the velocity vector); tighter lines above the humps indicate faster flow, looser lines under the humps indicate slower flow. **(b)** Localization of the vortex ω around the inflection point P on the upstream face of the interfacial perturbation.

Because the reference frame moves to the right with the average velocity of the two fluid layers, an observer watching the shearing between a fast wind moving to the right and a sea practically at rest sees the point P where the vorticity is concentrated as being on the upstream face of the surface wave generated by this perturbation. The slope of the upstream face will progressively increase, because the induced rotation tends to move point N above point P and point Q below point P. The upstream face of the wave thus becomes progressively shorter, whereas the downstream face becomes longer. If the shear is maintained, a limit is finally attained when the upstream face is vertical, which means it has infinite slope. This situation introduces us to a fundamental and very general property of the equations of fluid mechanics: in a perturbed regime, the faster parts necessarily catch up with the slower parts. This allows us to understand not only the breaking of waves in the open ocean but also the formation of shockwaves and hydraulic jumps.

This vortex, which is initially distributed continuously along the interface but that becomes ever more localized, is one of the characteristic properties of the KELVIN–HELMHOLTZ instability. It explains the formation of the quasi-periodic chains of small clouds that roll up into spirals near each inflection point analogous to point P and that are visible in the photograph in figures 4.5 and A.4. These spiral clouds, all turning in the same direction, are generated by the condensation related to the centripetal depression and can be considered as markers giving the location of shearing layers in the atmosphere. Moreover, because the mechanism behind the instability is based only on any eventual differences in density, the interfaces between different fluids are also subject to the KELVIN–HELMHOLTZ instability. Nevertheless, the influence of gravity, which stabilizes the system when the heavier fluid is below the lighter fluid, modifies the critical conditions. Thus, no sooner does a violent gust of wind destabilize the free surface of the seas and of lakes than these waves are generated. As they evolve, their upstream face rises up and until the wave breaks.

Figure A.4 - Example of a cloud layer deformed into spirals all turning in the same direction. These clouds indicate the presence of a shearing between an air layer above the clouds that moves to the right and the cloud layer itself, which moves more slowly. [© Terry ROBINSON/UCAR]

Unlike the RAYLEIGH–BÉNARD instability, the KELVIN–HELMHOLTZ instability has no critical threshold; instead, it has a critical wavelength. The vorticity may begin to localize for all perturbations that are longer than this critical wavelength. However, the wavelength corresponding to the distance between the first visible spiral vortices is always more amplified than the other wavelengths (fig. A.5a). As soon as they form, these vortices, which rotate in the same direction, begin to influence each other, leading to a remarkable pairing mechanism: n induces $n+1$ to pass below itself and $n+1$ induces n to pass above itself (fig. A.5a). This pair (fig. A.5b) is twice the size of the initial vortices and becomes progressively more homogeneous until it forms a single, larger vortex. Next, this scenario can iterate over and over until terminated by viscous friction. The image in figure A.4 is no doubt the ultimate, stabilized, stage of this scenario of successive pairings. Later, we will again encounter this tendency of vortices to pair up when we discuss the properties of two-dimensional turbulence.

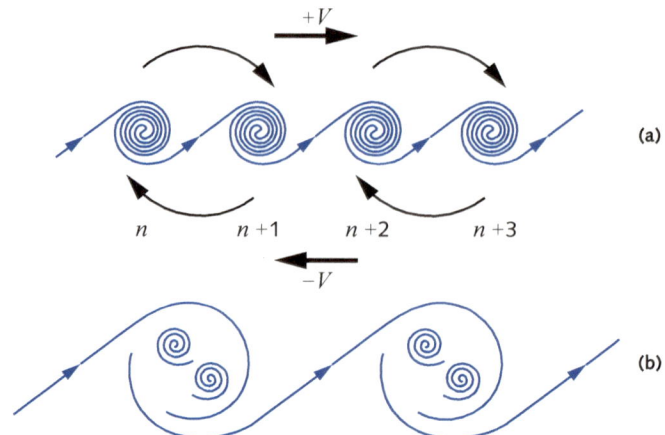

Figure A.5 - Schematic representation of the formation of vortices generated by the KELVIN–HELMHOLTZ instability. **(a)** The first vortices form due to the KELVIN–HELMHOLTZ instability and the first pairing occurs. **(b)** A new pair of vortices develop that can now partake in a new pairing.

Before closing this section, it seem useful to attract the reader's attention to the difference between the form of a perturbation near the instability threshold, which is characterized by a precise dimensionless number, and the final form of the perturbation, which we can observe in nature and where this dimensionless number is generally much greater. This final form is actually the result of an entire series of transitions that follow one upon the other starting with the initial perturbation. The theory that predicts the initial threshold is linear and therefore relatively simple; it cannot predict the amplitude of the final perturbation. More complex theoretical methods, essentially nonlinear, are required to analyze the path that leads to the successive perturbed states and eventually to the fully turbulent regime. The analysis of experiments also requires particularly refined diagnostic methods to follow the path of these successive transitions.

3. Other common vortex structures

The vortices that form via the KELVIN–HELMHOLTZ instability all rotate in the same direction, which is that of the initial vorticity of the shear layer. Rotating in the same direction means that the vorticities of all the vortices have the same sign and is a required condition for them to pair up. What happens to vortices of opposite sign? Figure A.6 schematically shows this situation to make it easier to understand: each eddy induces its partner to move at the same speed V, the end result of which is that both are propelled along in unison at the same speed. This configuration can continue over long distances without falling apart because only the viscosity and turbulence of the surrounding fluid can affect it, which is done by their combining to diffuse the vorticity out of the initial vortices. Given enough time to do its work, this double mixing and diffusion mechanism annihilates the two opposite vorticities and the vortex pair disappears.

Figure A.6 - Cross-sectional view of a pair of vortices A and B of opposite sign and with parallel axes (both axes are perpendicular to the page). These vortices mutually induce each other to translate at the same speed V because $V_{B/A} = V_{A/B}$. They do not wrap around each other as do those in figure A.5; this pair move in unison upward at speed V.

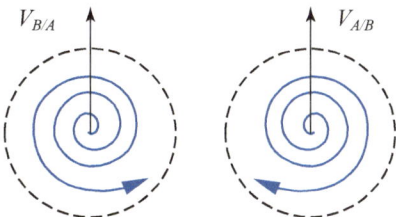

This condition is met more often in axisymmetric systems, such as smoke rings (fig. A.7). These self-propel themselves along the direction of their axis but over a rather short distance before diffusing their vorticity, which stops and destroys them. Other such axisymmetric structures include the turbulent tori that surround the mushroom-shaped plumes from explosions or those that form at the exit of an ejector such as a chimney or the crater of a volcano. All cross sections of tori taken in the plane of symmetry that contains the axis appear exactly as shown in figure A.6.

Figure A.7 - Smoke ring in the form of a torus self-propelling itself along the direction of its axis while growing and degrading, as shown by the traces of smoke left in its turbulent wake. [© Ali Momeni]

The flow of low-viscosity fluids, such as air and water, around obstacles leads to the formation of a rather peculiar and spectacular turbulent structure: a von Kármán vortex street[6] (fig. A.8). This structure forms in particular downstream of non-streamlined bodies. A cylinder, for example, works better than a thin wing to illustrate the origin of this phenomenon.

Above and below the cylinder in figure A.8 appear boundary layers. Their thickness grows gradually from extremely thin near the upstream stagnation point to rather thick beyond the points of maximum speed. From these points fluid filaments that

---

6   Theodore von Kármán (1881–1963) was a Hungarian engineer and physicist who became American during World War II. He was a specialist in aerodynamics and turbulence and, notably, founded the Jet Propulsion Laboratory (JPL) at the California Institute of Technology (Caltech).

breach the boundary layer and that are already slowed by friction no longer have sufficient momentum to fight against the pressure gradient that they meet on their way toward the downstream stagnation point. Thus, the boundary layers separate from the wall of the cylinder, creating there a negative speed. Downstream of the separation zone may appear a stable and symmetric regime, as illustrated in figure A.8a and which is characterized by an area of so-called *calm water* immediately downstream of the obstacle. However, as soon as inertial effects become more significant than viscous effects; in other words, when the REYNOLDS number,[7] which measures their ratio, becomes much greater than unity, this zone of calm water becomes unstable and, instead of the two stationary vortices, mobile vortices of alternating sign are periodically generated and carried away in the wake.

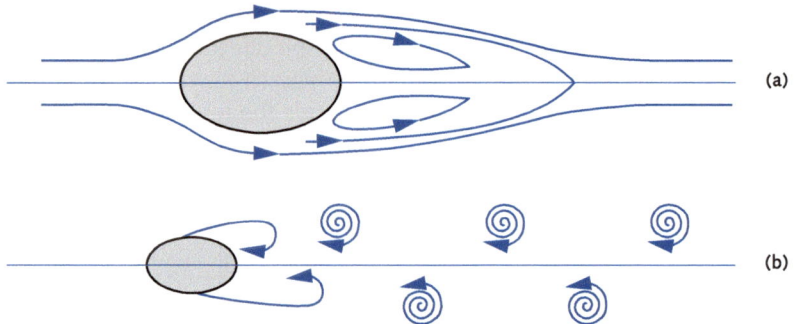

Figure A.8 - Illustration of instability in the flow around a cylinder. **(a)** A calm area of water appears downstream of the cylinder and remains stable for moderate REYNOLDS number ($Re \approx 10^2$). **(b)** Periodic releases of VON KÁRMÁN vortices of alternate sign when the REYNOLDS number is much larger.

The vorticity of the vortices shown in the top part of figure A.8b all have the same sign (negative) and those on the bottom have the opposite sign (positive). Despite being generated by the instability of the stationary regime shown in figure A.8a, this street of alternating and staggered vortices constitutes a remarkably stable state characterized by a well-defined STROUHAL number.[8] This frequency coordinates the periodic releases of vortices. The turbulence that coexists with these vortices diffuses their vorticity much more efficiently than can be done by viscosity alone and results in their rapid growth, as shown in figure A.9. However, this

---

7   The REYNOLDS number is expressed as $Re = VL/v$, where V and L are the typical speeds and lengths and v is the kinematic viscosity of the fluid. It can be interpreted as the ratio between the characteristic timescale of viscous friction $\tau_v = L^2/v$ to the timescale of transit $\tau_{tr} = L/V$, because $Re = \tau_v/\tau_{tr}$. For a boundary layer, the most pertinent length L is the thickness of the boundary layer, which is often written as δ. For flow around an obstacle, the most pertinent length is instead the characteristic size of the obstacle.

8   The frequency at which the vortices are released is such that, for a very large range of REYNOLDS numbers, the STROUHAL number $St = fD/V$ (where f is the frequency, D is the diameter of the circular cylinder, and V is the speed of the distant flow) is constant at 0.2. This result was discovered by Vincent STROUHAL (1850–1922), a Czech physicist and professor at the University of Prague.

mechanism is slow enough for the VON KÁRMÁN vortex street to be visible over several wavelengths.

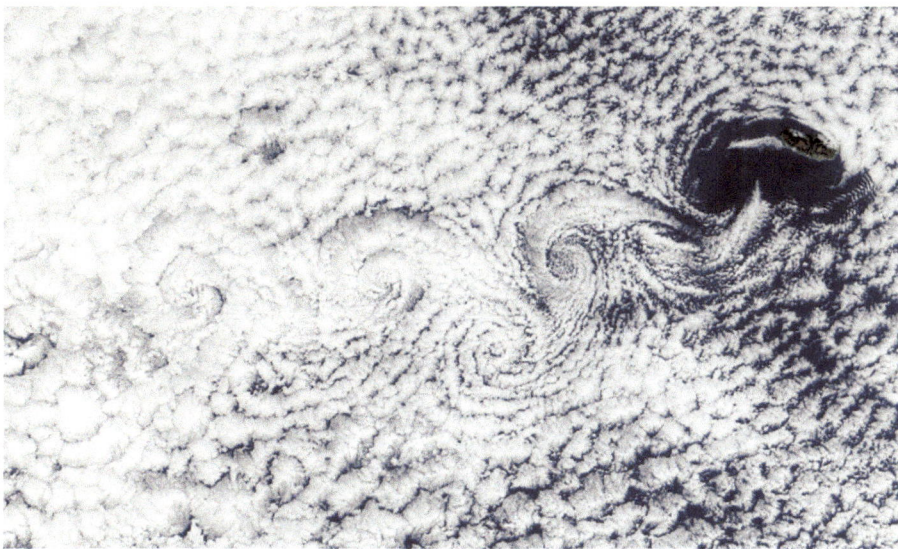

Figure A.9 - View from above of alternating VON KÁRMÁN vortices in a turbulent wind blowing clouds over Madeira Island. This structure is actually at an altitude much higher than the highest points of this island because of the inertial waves created by the topology of the island, which generate a TAYLOR column. The situation is essentially as if the wind encountered not just this island, but a sort of infinitely high vertical cylinder based on the island and separated from the surrounding wind by an intangible curtain. The wake of the cylinder thus forms at the altitude of the cloud tops, which are visible from space. The small clouds in the wind external to the wake reveal the characteristic length scale of the pressure fluctuations in turbulent flow such as this. The photograph was taken on December 1, 2002 by a NASA satellite. [© Jeff SCHMALTZ, MODIS Rapid Response Team, NASA/GSFC]

4. Transition toward turbulence

4.1. Appearance of turbulence in unconfined flows

Along a wall, such as the wing of an airplane flying through an infinite medium, we have seen that a boundary layer develops. Near the leading edge, this layer starts in a state that is perfectly reproducible and completely free of fluctuations. We call this state *laminar*. However, as soon as inertial effects overcome viscous effects, this layer becomes unstable, as can be seen in figure A.10 (this happens when the REYNOLDS number, which is based on the thickness of this boundary layer and which measures the ratio of these two effects, exceeds unity). The layer thickness then rapidly increases because the exchange of momentum between the slow layers in the immediate vicinity of the wall and the more distant layers are multiplied by the turbulence. Remarkably, the cross-sectional view of fluid flow around a wing shown in figure A.10 reveals that the transition to turbulence is accompanied by the formation of several large spiral vortices along the interface with the exterior laminar flow.

More precise observations of boundary layers around flat plates that do not fly show that this transition to turbulence passes through some preliminary and relatively well-organized stages, leading to the appearance of periodic longitudinal grooves spaced spanwise over the plate. Within this network of grooves, turbulence develops in a very sudden manner. Turbulent puffs form here and there and grow while being carried downstream where they form the turbulent layer. Some ruses, such as roughness or grooves machined into the plate, can advance or retard this transition, allowing the drag of a wing to be efficiently controlled.

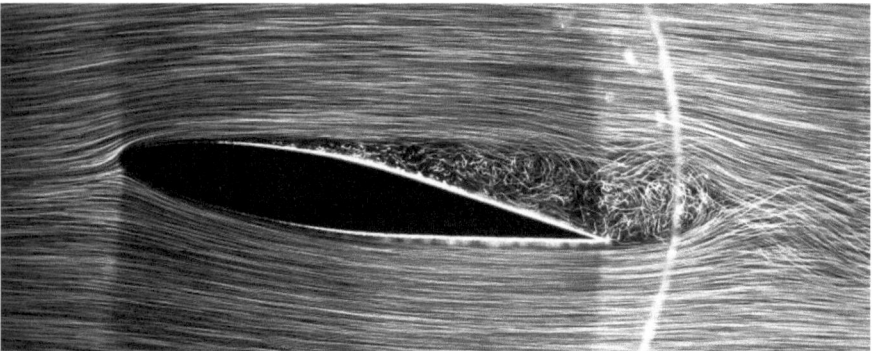

Figure A.10 - Visualization of flow around a wing with a nonzero angle of attack. Note the position of the stagnation point slightly below the leading edge, the point on the extrados downstream from the leading edge where the boundary layer separates from the wing, the almost immediate instability of the separated layer, and the formation of spiral vortices along the boundary with the exterior flow. [© ONERA, The French Aerospace Lab.]

Consider the smoke from the cigarette shown in figure A.11, which is not a boundary layer but rather a large-scale hot plume. Although this phenomenon pales in comparison with the wind and cyclones, since the REYNOLDS number is fairly small (of the order of several hundred), it has the merit of showing both the laminar regime and the transition toward the turbulent regime. The laminar ascension, in which the fluid filaments essentially do not mix, is maintained for some tens of centimeters. Next, we see a transition toward the turbulent regime, which forms some random vortices that cause the plume to suddenly become significantly wider. The plume is no longer static and the fluid filaments are mixed thoroughly. Successive photographs of this ascending smoke would show the relatively reproducible character of the laminar filaments and, in contrast, the extremely random character of the vortices that form after the instability.

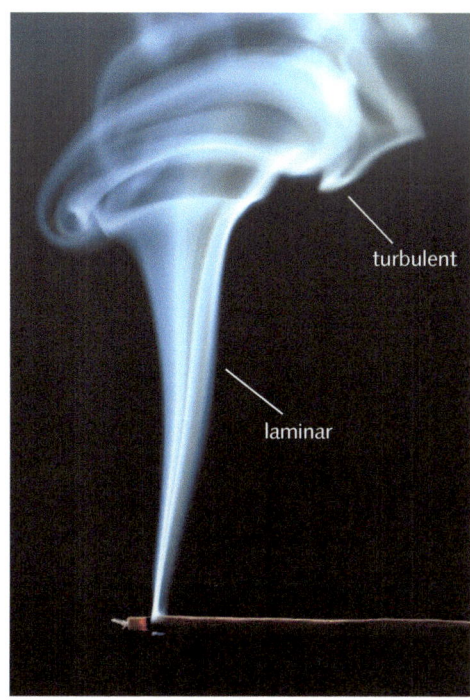

turbulent

laminar

Figure A.11 - About 10 cm from the initial point, the smoke from a cigarette transitions from the initial laminar flow to theturbulent regime.[© Freeimages.com/ Krunoslav Knežević]

4.2. Transition toward turbulence in duct flows

Ducts with circular cross sections are extremely common in hydraulic facilities, notably those discussed in chapter 8 in connection with large waterworks. The flow through such ducts is profoundly affected by the confinement of the fluid imposed by the walls of the duct, which translates notably into the conservation of flow rate all along the duct and by a particularly subtle laminar-turbulent transition. As far back as 1880, Osborne Reynolds himself observed with the aid of visualization methods that, in a very long duct, turbulent puffs appeared suddenly at random points when the ratio VD/v, which is now named after him, was sufficiently large. These turbulent zones were transported downstream at a speed slower than the speed of the fluid in the duct and were separated from each other by laminar zones. We shall revisit later the critical magnitude of this Reynolds number VD/v, where V is the average speed of the flow in the duct (flow rate divided by the cross-sectional area), D is the duct diameter, and v is the kinematic viscosity of the fluid. About a century passed before anyone noticed that fluid particles moved through these puffs, exiting from the downstream side without carry away any turbulence and without contaminating the downstream fluid. The word *intermittency* was introduced to designate this curiosity, which has since been found to be one of the profound characteristics of this transition to turbulence in confined flows.

The particularity of this transition comes from the fact that the threshold at which it appears is not defined only by the REYNOLDS number. It also depends on other parameters that are relatively difficult to discern, such as details of the conditions under which the fluid enters the duct, any eventual external vibrations, perturbations from the pump or ventilator, or the roughness of the wall of the duct. In hydraulic facilities, if no particular precautions are taken, the transition occurs as soon as the REYNOLDS number exceeds approximately 2400. But in the laboratory, depending on the precautions taken, this critical value can grow much larger, up to around 10^5 or 10^6.

The differences in the critical value are explained by the fact that turbulence appears in this flow not because of a transition from an initially infinitesimally small perturbation, as is the case for the RAYLEIGH–BÉNARD instability. On the contrary, the laminar regime of this POISEUILLE flow is systematically stable against infinitesimal perturbations. This flow is called *linearly stable*, in reference to the linear methods used to analyze unconfined flows and that lead in particular to the RAYLEIGH–BÉNARD and RAYLEIGH–TAYLOR instability thresholds. In contrast, POISEUILLE flow is unstable because it is subjected to external perturbations of finite amplitude, and the critical REYNOLDS number depends on the properties and the amplitude of these perturbations.

In the industrial facilities described in chapter 8, the effective REYNOLDS numbers are so large, in the range of 10^5 to 10^7, that turbulence is always present. It can nonetheless evolve as we observe it a short distance downstream from an external perturbation that generates additional turbulent energy, such as a bend in the duct, a partially opened valve, or a change in the duct cross section. However, after a relatively long, straight section, the turbulence may reach its asymptotic and stable state. In this situation, far downstream of any localized excitation, the properties of turbulence, such as its kinetic energy and the way it is partitioned between the three velocity components, do not vary along the abscissa. Thus, for each unit length along the duct, the energy withdrawn by turbulence from the average motion compensates exactly for the turbulence energy dissipated by viscosity.

4.3. Other ways to generate turbulence

Diverse remarkable transitions occur, both in the atmosphere and in the oceans and rivers, where we find the characteristics evoked in the preceding sections. Figure A.9 shows an example of very-large-scale turbulent flow in the atmosphere, where the REYNOLDS number is no doubt greater than 10^{10}. The TAYLOR column (see insert I2.5) formed above Madeira Island acts like an obstacle to the wind coming from the right; it generates in its wake a VON KÁRMÁN alternating-vortex street, which is the organized part of the flow and which coexists with turbulence on smaller scales that is already present in the upstream wind. Downstream of the island, the energy of the large alternating vortices induced by the TAYLOR column is progressively transferred to the turbulence and contributes to its reinforcement.

This is the situation for all perturbations of a turbulent wind or marine current, be they natural such as atmospheric depressions and cyclones, or generated by human activity, such as the wakes of airplanes or ships. Atmospheric or marine turbulence, such as those evoked in this appendix and in many others as well, must thus be viewed as the result of complex and very diverse excitations that act at diverse locations. These disturbances influence the flow over long distances downstream from where they are generated; they slowly attenuate and disappear so that, where turbulence occurs, it has already accumulated the energy supplied by various actions. Turbulence combines all the external influences, gradually ridding itself of the specific characteristics of each and developing instead its own dynamics. Far from any excitation, the turbulence ends up in a sort of universal equilibrium state, which is analyzed in the following section.

5. Fully developed turbulence

5.1. Turbulence in the most common flows

In wind tunnels and laboratory water channels, flows that are not subjected to any force with specific properties, far above the critical conditions, manifest a well-developed turbulent regime. Even if certain properties still raise challenging questions that we shall not discuss here, this turbulence is now sufficiently well understood that is must be considered as a reference before examining any eventual special conditions, notably in the atmosphere.

One of the essential characteristics of well-developed turbulence is, for very large REYNOLDS numbers, the large range of its energy spectrum. Large vortices, which carry the largest fraction of the total kinetic energy, generally have a size that is adapted to the length scale of the mechanism that created them, such as the diameter of a duct, or the width of a wake. The smallest size, however, is imposed by viscosity; in a fluid such as air this size is of the order of hundredths of millimeters. The energy supplied to the largest vortices by the mechanism that creates them is transferred to ever-smaller structures and thus travels over the entire energy spectrum in a so-called *direct cascade process*,[9] as already evoked. Note the relatively passive role of viscosity, which transforms kinetic energy into heat and, in so doing, selects only the scale at which it may act and at which the energy cascade stops.

From the point of view of its practical consequences, this turbulence very efficiently mixes all quantities that are susceptible to being transported, such as heat or the concentration of a particular chemical species. The range of the exchanges, which

---

9   The name of Andreï KOLMOGOROFF (1903–1987), a Russian mathematician and professor at the University of Moscow, is the name attached to this energy cascade. He discovered one of its universal characteristics; namely, a spectral energy distribution that follows $k^{-5/3}$, where $k = 2\pi/\lambda$ is the wave number, with λ being the wavelength.

would be the mean free path of molecules in laminarly flowing air, becomes the size of the most energetic vortices; that is, the largest. In the same way, the typical speed of fluid particles in this turbulence replaces the speed of the molecules. Let us take, for example, an atmospheric turbulence whose most-energetic vortices have a size $l \approx 10$ m and a fluctuation in speed $u \approx 0.1$ m s^{-1}. To evaluate the effective viscosity, we must replace the product $c\lambda \approx 10^{-4}$ m^2 s^{-1} introduced in chapter 1 by the product $ul \approx 1$ m^2 s^{-1} (see insert I1.5). We can thus understand that the effective viscosity due to turbulence, as for the apparent thermal conductivity and all other transport properties, is multiplied by a very large factor (of the order of 10^4 in the present case of atmospheric turbulence). For the flow of water through a duct, we obtain a smaller order of magnitude, in the neighborhood of 10^3, although this is still much greater than unity. These estimates, although approximate, suffice to show the incredible capacity of turbulence to amplify viscous friction and to improve mixing.

5.2. Large-scale atmospheric turbulence

Whereas small-scale turbulent motion, on the scale of, say, tens of centimeters or less, has properties very close to those of a uniform-density fluid, as summarized in the preceding section, the situation is quite different for atmospheric flows, which can possess horizontal scales much longer than tens of meters. To begin, consider length scales between 10 m and 10 km, which we shall call *intermediate*. Their characteristic lifetime (less than an hour) is much less than a day, the CORIOLIS force is negligible, but gravity and variations in density are sufficient to lead to a significant stratification.[10] The characteristics of turbulent structures of this intermediate size depart from those of a turbulence that is not subjected to an exterior force. The vertical speed, confronted with gravity for a sufficiently long period, cannot develop normally and remains much slower than the horizontal speeds. These structures thus acquire a flattened form and have no choice but to spread horizontally like a sort of disc or pancake. They can be stacked one on top of the other, while retaining the possibility of sliding with respect to each other. Their flattening strongly diminishes their capacity to mix the contaminants within the various horizontal planes but maintains their large apparent diffusivity in the horizontal directions.

The mechanism responsible for this stratification effect on intermediate scales in atmospheric turbulence is analogous to that which deviates a smoke plume. Initially vertical, a smoke plume curves rather suddenly and becomes horizontal when the next-higher layer of air is lighter than the plume itself, which can thus not penetrate it. This horizontal deviation is illustrated in figure A.12, where we

---

10 As its name indicates, stratification of a fluid consists of forming successive layers whose density decreases with altitude. This stack of layers is stable because, if one element is lifted into a lighter layer, gravity tends to return it to its original position. Inversely, a lighter element that falls into a heavier fluid layer is also lifted back up to its initial position.

note that the typical diameter of the plume is perhaps on the order of several hundred meters, similar to the diameter of the volcano, whereas the billows of smoke within the plume belong rather to the small three-dimensional scales discussed in the preceding section.

Figure A.12 - Example of three-dimensional turbulence in the eddies of smoke from a volcanic eruption. Note the overall horizontal expansion of the cloud above a certain altitude, which is due to stratification of the atmosphere. Note also the three-dimensional character of the eddies. [© R. RUSSEL, Alaska Division of Fish and Game/USGS]

In contrast, for large-scale flows at least on the order of 100 km, such as for atmospheric depressions, the CORIOLIS force is no longer negligible with respect to stratification. A priori, these two effects share a characteristic—the vertical direction. But they also have an important difference: families of inertial waves exist in rotating systems (see insert I2.5) and propagate along the axis of rotation, tending to homogenize the speeds in this direction. This mechanism, which is superposed onto that of stratification, has important consequences: on scales sufficiently large for the CORIOLIS force to dominate, inertial waves establish a good correlation between the diverse horizontal planes, tending to form vertical columns instead of separate discs. If we compare vortices flattened by stratification with discs or CDs stacked one on top of the other, the inertial waves have an effect analogous to that of the central rod that prevents the discs from sliding over each other. Columnar structures thus formed are visible in the laboratory in models that have similarity with the large-scale atmospheric effects. These are illustrated in figure A.13.

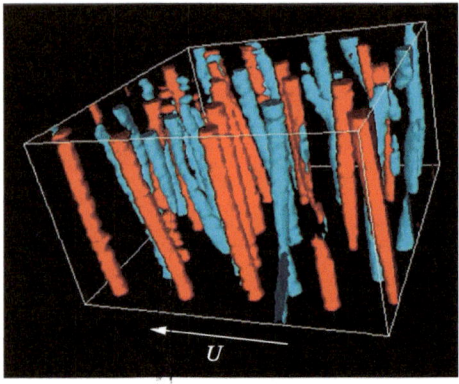

Figure A.13 - Experimental demonstration of the formation of quasi-two-dimensional columns in a rotating stratified fluid. The red and blue surfaces have the same magnitude vorticity but of opposite sign. The turbulence is generated by a grid positioned to the right of the figure (not shown) and through which the fluid flows horizontally at speed U.
[© Joël SOMMERIA, Olivier PRAUD and Adam FINCHAM/LEGI, CORIOLIS, Grenoble]

Let us retain that atmospheric turbulence simultaneously possesses very large scales characterized by a tendency to become two-dimensional and to be well structured in vertical columns, intermediate scales that already have a certain quasi-two-dimensional characteristic related to their stratification but which do not form columns, and finally the smallest scales that remain three dimensional. Anticipating the next paragraph, we can already say that the large scales rapidly eliminate the anticyclonic vortices, leaving only the cyclonic vortices. These tend to pair up and thus form an inverse energy cascade toward ever larger scales. The small scales, on the contrary, are three dimensional and form a direct energy cascade that results in their energy being dissipated by viscous friction. The influence of the small and intermediate scales (three dimensional) on the largest scales (quasi-two-dimensional) can, in a first approximation, be reduced to a super viscosity or effective viscosity, which may be at least one thousand times stronger than the actual viscosity, whose origin resides in molecular movements.

We have already remarked that the REYNOLDS number for atmospheric flows can be extremely large (of the order of 10^{10} or more). On the scale of the turbulent fluctuations themselves, this is still very large. Consider the example of a small three-dimensional scale: a vortex of size $l \approx 10$ m and within which the difference in speed with the surrounding wind is of the order of 1 m s^{-1}. The REYNOLDS number of this vortex is about 10^5, which means that the characteristic timescale for a revolution of the vortex is of the order of $l/u \approx 10$ s, which is 10^5 times shorter than the time required for the viscosity to dissipate its energy. The conclusion is clear: the dynamics of the vortex is essentially independent of the viscosity and is instead purely inertial.

5.3. Inverse energy cascade
in two-dimensional turbulence

Beyond time comparable to that of a single revolution, all conceivable vortex structures would have interacted with their neighbors to form other such structures. The interactions between vortex structures offer two possibilities.[11] They can assemble to form larger structures; a possibility we have already noted for the pairings between spiral vortices generated by the KELVIN–HELMHOLTZ instability and which forms an inverse energy cascade. But they can also divide to form smaller vortices, giving rise to a direct energy cascade as already evoked. The most common examples of this direct energy cascade are no doubt the winds and the flow of water in streams, phenomena on the scale of our senses and with which we are very familiar. But how does nature chose between these two possibilities and how do they differ?

The direct cascade toward smaller scales is energetically coherent from all viewpoints: production of large vortices by an instability, and then cascade toward smaller scales leading, at the end of a continuous spectrum of energy but always at the same point in space, to small vortices whose energy is dissipated by viscosity. We can convince ourselves by comparing the time required for the viscosity to dissipate a 10^{-1} mm vortex, which is of the order of 10^{-3} s, with its own period of revolution, which is also of the order of 10^{-3} s if it contains fluctuations in speed of the order of 10^{-1} m s^{-1}. Clearly, a smaller vortex would not have enough time to make a single revolution before its energy is dissipated into heat. We can detect the presence of these structures, which are small enough for the turbulence to remain three dimensional, in the core of a jet, on the downstream side of figure 4.7. This figure also reveals the presence of spiral vortices that form at the exit of the orifice and that do not have sufficient time to pair up before they transfer their energy to the turbulence of the jet.

The inverse cascade toward ever larger scales is only visible in quasi-two-dimensional systems, where the special direction is imposed by an outside force. We shall

---

11 Let us schematically describe vortices of size l_1 and l_2 by sinusoidal speed distributions in x: $\sin p$ and $\sin q$, where $p = \pi x/l_1$ and $q = \pi x/l_2$. Because their interaction is purely inertial, which is to say quadratic, the resulting vortex must correspond to the speed distribution given by the product 2 $\sin p \sin q = \cos(p-q) - \cos(p+q)$. We can thus understand how, starting with structures of size l_1 and l_2, larger or even much larger structures [of the order of $l_1 l_2/(l_2-l_1)$] or much smaller structures [of the order of $l_1 l_2/(l_2+l_1)$] can form.

focus on very large atmospheric depressions because they possess this property. However, this inverse cascade raises several questions. First, if the average vorticity is zero at a certain instant (because an equal amount of opposite-signed vortices were created, as in the case of a VON KÁRMÁN vortex street shown in figure A.8b), how is it possible that the interactions give rise at a later time to a state with a non-zero vorticity? Two mechanisms compete to reinforce the vorticity of cyclones (i.e., a vortex that, were it slid up to the North Pole, would rotate in the same direction as Earth) and to deteriorate the vorticity of anticyclones.

The first mechanism is the contribution of the CORIOLIS force, which reinforces the cyclonic vorticity in the Northern Hemisphere and the anticyclonic vorticity in the Southern Hemisphere. To explain this mechanism in a simplified manner, consider the motion of a vortex with angular speed ω due to the rotation of a global reference frame rotating at angular speed Ω, as represented in figure A.14. During the time t, the two rotations ωt (rotation of the cross $ABCD$ drawn on the cyclonic vortex) and Ωt (rotation of the reference frame) add together, so that the cyclonic vortex is endowed with strong angular momentum in the absolute reference frame. Because this quantity is essentially constant, the lifetime of the vortex is very long. For a vortex of the opposite sign, the two rotations counteract each other, thus minimizing the angular momentum and the lifetime of the structure.

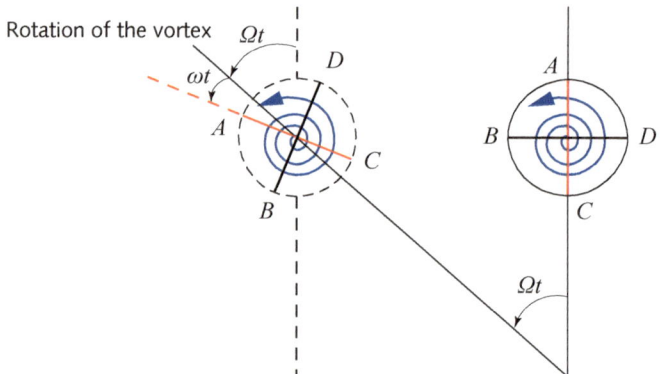

Figure A.14 - Schematic diagram showing evolution of a cyclonic vortex rotating with angular speed ω in a reference frame (Earth) rotating with angular speed Ω. During the time t, the rotation Ωt of the reference frame is added to the rotation ωt of the vortex, which reinforces the cyclonic vortex.

The second mechanism acts even on small scales but requires the presence of a medium vorticity of a given sign, as for a shear layer subjected to a KELVIN–HELMHOLTZ instability. We saw above in this appendix (fig. A.2 and A.3) how this instability first localizes the vorticity and then enables the growth of spiral vortices (fig. A.5). Imagine now the presence of a vortex of the opposite sign in between two spiral vortices that are in the process of pairing. The three panels of figure A.15 show that the speed induced by the pair of vortices with the same sign stretches more and more the weaker, opposite-signed vortex, thinning it so that it takes the form of a filament or a sheet that wraps around the pair of cyclonic vortices. During this process, the filament or sheet becomes thinner and thinner. At a certain point, it becomes thin enough for viscosity to do its work of dissipation, making this opposite-signed vortex disappear.

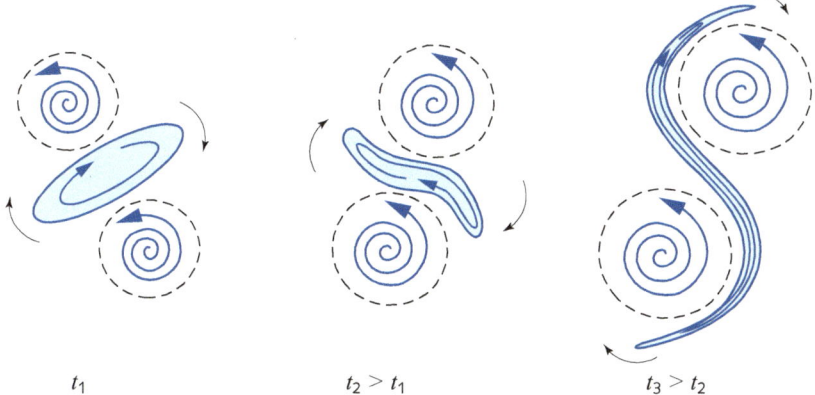

t_1 $t_2 > t_1$ $t_3 > t_2$

Figure A.15 - Action of two vortices with the same sign and in the process of pairing on a third vortex of opposite sign. At successive times t_1, t_2, t_3 the anticyclonic vortex (blue) becomes ever more stretched and, due to conservation of mass, become ever thinner, to the point that viscosity can dissipate its kinetic energy into heat.

At this stage, to confirm the energy coherence of the process, an important question again arises: because viscosity has no effect on such large scales, how is the energy constantly supplied by the inverse energy cascade dissipated into heat? The answer comes from the atmospheric boundary layer, or EKMAN layer, which, over large horizontal scales, subjects these columnar structures to friction on the ground and on the seas. This aspect was described and commented in chapter 2.

Conclusion

The two fluids of interest in this work, air and water, exhibit permanent agitation around what we often call a *mean flow* by combining multiple, apparently disorganized fluctuations that are nonreproducible except in their overall properties. This agitation has diverse origins that are related to various families of instabilities, the most important of which, for these natural media, were described in the first sections of this appendix. Via complex paths that we have not detailed, the kinetic energy thus withdrawn from the unperturbed part of the motion supplies energy for the turbulence. This fundamentally random and nonreproducible phenomenon remains mysterious in many ways, although its main properties were shown and can be simulated now with rather precise numerical techniques. We emphasized the fact that turbulence cannot be reduced to a small perturbation superimposed over an average flow; on the contrary, it strongly affects the flows and distributions of all transported quantities, such as heat and chemical species in solution.

Atmospheric turbulence possesses both a three-dimensional dynamics that dominates on the small scales, and a two-dimensional dynamics that dominates on the larger scales. The specialists know how to recognize and distinguish their specific mechanisms, particularly the direct energy cascade on the small scales and the inverse energy cascade on the larger scales. The latter turns out to be essential for understanding the evolution of the large atmospheric depressions, which are rendered two dimensional by inertial waves, themselves generated by the CORIOLIS force. These large scales are thus forced to transfer their energy to even-larger structures, until reaching a limit imposed by the atmosphere, where friction with the ground finishes by transforming this kinetic energy into heat.

If we have not specifically considered the case of the seas and rivers in this appendix but instead cited examples involving the atmosphere, it is because they do not differ much from common conditions. In rivers, turbulence is three dimensional; its energy is supplied by the wakes of all the obstacles to the flow, by the meanders, and by hydraulic jumps and it is subject to a direct energy cascade toward the smaller scales. In the seas, the novelty is related in part to its interaction with waves, which turbulence slows by removing some of their energy, and by the thermocline, which constrains turbulence to the first tens of meters and prevents its penetration into the depths.

The reader will have noted the major role of the planetary heat engine, where sunshine is the unique source of temperature differences and of exchanges of heat. This generates the main movements, from the trade winds on a planetary scale to the local thermal winds, not forgetting the gigantic thermohaline circulation. Turbulence can be considered as the ultimate and inevitable stage of the effects of this sunshine, because the smallest scales with their fine shearing are where the kinetic energy of the aerial and marine motions are finally converted to heat by using viscosity, as weak as it may be.

Glossary

Acoustic waves - These waves propagate in matter via the interactions between elementary particles, atoms, or molecules, which transmit from neighbor to neighbor all perturbations that they receive. Such waves are also called *sound waves*. When these waves have frequencies ranging from 15 to 20000 Hz, they can be detected by human hearing.

Aerosols - These consist of solid dusts or liquid droplets smaller than 10 μm in size and that remain suspended in the air without precipitating because of their weight, which is small with respect to the forces exerted on their surface. Aerosols serve as seeds around which water molecules can aggregate to form either ever-larger drops, which lead to rain, or flakes that lead to snow. Aerosols have a very weak cooling influence on the climate. The 0.3 W m^{-2} of cooling that they provide is essentially negligible compared with the greenhouse effect, which contributes 150 W m^{-2} of warming.

Albedo - The fraction of solar radiance directed toward Earth that is reflected toward space and does not participate warming the planet. This reflection is essentially due to the large masses of water in the atmosphere, such as clouds, or at the surface of the planet, such as snow-covered ground, glaciers, and oceans.

Anabatic wind - This is the wind that, in mountainous regions, blows up the slopes in the afternoon once the high-altitude regions have been well warmed. The warmed and thus lightened air rises, drawing up after the air from the valleys. It is the inverse of the katabatic wind (see definition).

Angular momentum - A mass m at a distance r from an axis Δ about which it moves in circular motion with angular speed ω possesses a momentum $m\omega r \mathbf{e}_\theta$, where $\mathbf{e}_\theta$ is a unit vector, is parallel to Δ, and indicates the direction of rotation about Δ. Mass m also possesses angular momentum $\mathbf{L}_\Delta = m\omega r^2 \mathbf{e}_r \times \mathbf{e}_\theta = m\omega r^2 \mathbf{e}_z$, which has the important property of remaining constant in the absence of torques applied from the exterior.

Anticyclone (high-pressure zone) - A region of the troposphere where the pressure is greater than the normal pressure (1013 hPa at sea level). The pressure in a high-pressure zone can reach about 1080 hPa at sea level. High-pressure zones, which are surrounded by low-pressure zones (atmospheric depressions) and their accompanying winds, are very calm regions that move according to varying influences, notably that of the polar jet stream in the Northern Hemisphere.

© Springer International Publishing AG 2017
R. Moreau, *Air and Water*,
DOI 10.1007/978-3-319-65215-3

BOND number - The deformations of an interface between two nonmiscible fluids, such as air and water, strongly depend on the BOND number, which is the ratio $Bo = gL^2\Delta\rho/\sigma$. It is constructed from gravity g, the length scale L that characterizes the phenomenon in question, the difference $\Delta\rho$ in density between the two fluids, and the surface tension σ. Its value serves to compare the relative contributions of weight and surface tension to the evolution of the interface.

Boundary layer - In flow where the REYNOLDS number is much greater than unity, a boundary layer is always present near the surface of obstacles. This thin layer separates the surface itself from the distant flow, which is often irrotational, and is the only region where viscosity can balance the forces of inertia and pressure.

Capillarity - A phenomenon whereby water rises via so-called *capillary ascension* through small-diameter vertical tubes, such as in blotters, sponges, soil, and, more generally, in all porous media. Capillarity is due to the fact that water molecules wet fairly well solid walls and tends to climb along their surface rather than descending to avoid the wall, as is the case with mercury. Two properties control this phenomenon: the surface tension at the interface between two fluids and the contact angle of this interface with the wall (see insert 13.2).

Celerity (of waves) - A wave, no matter its nature, transports energy and, eventually, other information, such as its form in the case of a wave on the free surface of a liquid. The celerity of a wave designates the speed with respect to the medium in which energy is transported; it often differs greatly from the speed of the matter in the medium itself. In wave theory, two vector velocities are distinguished: a phase velocity, which is specific to each mode or each wavelength, and a group velocity, which characterizes the overall transport of energy by all the modes. The celerity is the magnitude of the group velocity.

Centrifugal force - Any mass moving along a curved trajectory must accelerate toward the interior of the curve. In the accelerating reference frame of the mass, the acceleration appears to be toward the exterior of the curve (e.g., the push you feel toward the exterior of the curve in a turning car). The magnitude of the acceleration is $\omega^2 r$, where ω is the instantaneous angular speed of the mass and r, is the distance from the center of the curve. If V is the tangential speed along the curve, ω is given by the simple formula $\omega = V/r$. Instead of speaking of an acceleration, the normal parlance refers to a centrifugal force so as to reflect the sensation described above of a passenger in a car rounding a curve.

Coalescence - A phenomenon whereby two separate nearby drops within another continuous fluid (a gas such as air, or another liquid) or two bubbles in a liquid join together to form a single drop or bubble. In so doing, the energy stored in the interfacial surface diminishes, as does the surface area itself, making the system more stable.

Cold front - This is the opposite of a warm front: a mass of cold, heavy air is forced to pass under a mass of warmer and lighter air, thereby lifting the warmer air. As in the case of warm fronts (see definition), the region at the interface between these masses of air can develop thunderstorms and precipitation.

Condensation - This change in phase is the opposite of evaporation: molecules in the gaseous phase, where they are very far from each other (see insert I1.3), enter the liquid phase where, without losing their disordered motion, they pack much closer together. This is reflected in the fact that a liquid occupies a given volume whereas a gas occupies the entire volume available. Condensation warms the surrounding medium by liberating the latent heat of the gas.

Convection - Fluids in motion are able to constantly renew the fraction of fluid near a heat source. Being at a lower temperature, a freshly renewed fraction is more rapidly heated by conduction than the warmer fraction it has just replaced. Continuing the cycle, the newly heated fraction is replaced by a cooler fraction, which will be heated and evacuated in turn. The transfer of heat is thus accelerated by the flow of the fluid. This additional contribution to the heat transfer is called *convection*.

CORIOLIS force - To describe relative motion in a rotating system, the rules for summing accelerations is more complex that for velocities, which is given by the vector addition $\mathbf{V}_a = \mathbf{V}_e + \mathbf{V}_r$, where $\mathbf{V}_a$ is the absolute velocity of the object, $\mathbf{V}_e$ is the velocity of the rotating frame of reference, and $\mathbf{V}_r$ is the velocity of the object relative to the frame of reference. For accelerations, the following formula must be applied: $\gamma_a = \gamma_e + \gamma_r + \gamma_c$, where, in addition to the acceleration γ_e of the reference frame and the acceleration γ_r relative to the reference frame, another acceleration appears called the *CORIOLIS acceleration* γ_c. As is done with the centrifugal force, the equation of motion $\gamma = \mathbf{F}/m$ is typically used to express the CORIOLIS acceleration in the form of a force per unit mass: $-2\Omega \times \mathbf{V}_r$, which is called the *CORIOLIS force*. The main effect of this force is that motion along a line for an observer in the inertial frame of reference appears to be curved for a terrestrial observer. In the Northern (Southern) Hemisphere the motion appears to curve to the right (left). This results simply because the planet constantly moves to the left (right) for the oxbserver in the Northern (Southern) Hemisphere.

Cyclone - In the troposphere, we observe alternating sequences of high- and low-pressure zones. When the latter reach a certain strength, we call them *cyclones* and give them a name and a number that indicates their strength. Remarkably, the CORIOLIS force obliges cyclones to rotate in the same direction as the planet (if you imagine sliding them to the North Pole, they rotate in the same direction as Earth). This direction of rotation is called *cyclonic* and is counterclockwise when viewed from space. Note that cyclones in the Southern Hemisphere seem to rotate in the opposite direction as seen by local terrestrial observers; however, this is due to the fact that they are upside-down with respect to the Northern Hemisphere.

Dendritic - This refers to a structure in the process of solidification that takes the form of a finger, often called a *digitation* or a *dendritic structure*, which protrudes into a fluid phase (liquid or gaseous). These dendrites often have secondary branches and result from the MULLINS–SEKERKA instability at the interface, which initially is plane or very slightly curved. The morphology of the solid-liquid interface formed from very closely spaced dendrites separated by liquid channels is extremely complex. Overall, the ensemble resembles a sort of miniature bouquet of fir trees

Depression - The air pressure in the troposphere varies around a mean value that is close to the so-called *normal pressure*: 1013 hPa at sea level. The term *depression* means a region where the pressure is less than the mean pressure and where the saturation vapor pressure is also lower than in the external region. If the air in a depression is loaded with moisture, it condenses and forms fog and clouds.

Dispersion - In almost all media traversed by waves, the different frequencies or modes do not propagate at the same speed, which leads to their separation or dispersal. The origin of this phenomenon occurs on the scale of the elementary particles that constitute the medium, and such media are called *dispersive*. The various modes in air that make up sunlight thus undergo dispersion due to the molecules in the atmosphere; notably, nitrogen, oxygen, and water. The color of the daytime sky, which ranges from indigo to light blue, is due to the moisture content of the air.

Eddy - This word refers to a rotating fluid structure that is poorly identified because its boundaries are not precisely defined, but to which we can attribute a form, a lifetime, and characteristic scales. Note that it differs from a vortex, whose vorticity is a quantity that is precisely defined at all points in a fluid whose motion is characterized by the velocity vector $\mathbf{V}$ in such a way that $\boldsymbol{\omega} = (\nabla \times \mathbf{V})/2$. The magnitude of the vector $\boldsymbol{\omega}$ is the local angular velocity of the rotating fluid particles.

EKMAN layer - In fluids rotating sufficiently fast for the CORIOLIS acceleration to dominate over other components of acceleration, the only region where this acceleration can be balanced by viscous friction is a very particular boundary layer called the *EKMAN layer* that spans from the ground to the high-altitude wind. The thickness of this layer is of the order of $\sqrt{v/\omega}$, where v is the kinematic viscosity of the fluid and ω is the angular speed of rotation about the local vertical. In the presence of turbulence, a correct estimate the thickness of the EKMAN layer requires using an effective viscosity instead of the much smaller intrinsic molecular viscosity. One of the specificities of the EKMAN layer is the continuous variation in the direction of the velocity vector: in the terrestrial atmosphere its direction at ground level is oriented at 45° with respect to the direction of the high-altitude wind (see fig. 2.9).

EKMAN spiral - In the EKMAN layer of the atmosphere, the distribution of speed as a function of distance from the ground involves both a variation in speed, which goes from zero at the ground to a maximum at altitude, and a variation in direction (see fig. 2.9). The EKMAN spiral refers to the projection onto the ground of this three-dimensional distribution of speed.

Electromagnetic waves - These waves represent the electromagnetic radiation from objects. They imply that the electric and magnetic field oscillate between each other, as described by MAXWELL's equations. Visible light is one of the most common natural manifestations of these waves.

ELVES - ELVES is an acronym that stands for Emission of Light and Very-low-frequency perturbations due to Electromagnetic pulse Sources. These red halos occur during storms and last less than a millisecond. They are visible in the stratosphere from space (see fig. 3.9).

Energy cascade - This expression is used in the theory of turbulence to designate the transfer of kinetic energy from vortices of a certain size toward vortices of slightly different size. This transfer, which is due to inertial effects, proceeds from neighbor to neighbor toward an asymptotic limit of zero or infinite size. When the energy cascade proceeds toward very small vortices, as happens in three-dimensional cases, it is called *direct*. It is called *inverse* when energy is transferred toward very large vortices, which happens in two-dimensional cases.

Evaporation - This phenomenon is the opposite of condensation (see definition): it is the transition from a liquid state to a vapor state. All fluids that undergo evaporation absorb a latent heat that must be supplied from an external source.

Fluvial or torrential regime - In a rectilinear waterway or stream, two regimes of uniform flow are possible. In one regime, the speed of the water is less than the celerity of the surface waves; this is called the *fluvial regime*. The depth in this case is rather large. In the other regime the speed of the water is greater than the celerity of the surface waves; this is called the *torrential regime*. In this case, the water is not deep.

FROUDE number - The respective contribution of inertial effects and gravity to the evolution of the flow of water at a free surface is characterized by the FROUDE number $Fr = V/\sqrt{gh}$. This number is proportional to the ratio between the speed V of the water and the maximum speed of an object in free fall from a height h, when its original potential energy ρgh is completely converted to kinetic energy $\rho V^2/2$.

Funnel cloud - When a tornado forms underneath a thundercloud, the funnel cloud is the vertical rotating tube that is darkened by its heavy charge of water drops and that descends from the bottom of the thundercloud toward the ground. The axis of rotation of the tornado is inside the funnel cloud, and the vorticity and centripetal depression are strongest there, which explains why it is so heavily loaded with the water drops that give it its dark color. Note that the surrounding air is also subject to circular winds whose angular speed decreases in inverse proportion to the distance from the axis.

Geoid - The gravitational equipotential surface that coincides best with the mean level of the seas (see fig. 5.3).

Geostrophic equilibrium - In fluids that rotate sufficiently fast for the CORIOLIS acceleration to dominate over components of acceleration, the fundamental equation of mechanics that applies outside of boundary layers imposes a balance between the forces of pressure and the CORIOLIS force. This equilibrium is called *geostrophic*. It leads to a curious property which is contrary to the usual intuition based on balances between forces where the CORIOLIS force is negligible: the velocity of the fluid must be perpendicular (as opposed to collinear) to the pressure force that drives the flow. This result is due to the fact that the CORIOLIS acceleration $2\omega \times V$ (where ω is the angular velocity of rotation about a local vertical axis and V is the velocity of the fluid) is perpendicular to the velocity V.

Greenhouse effect - Garden greenhouses are essentially transparent to daylight coming from the exterior and thus allow photosynthesis to proceed in the plants protected in the interior. In contrast, they absorb a large fraction of the infra-red radiation emitted by the ground and the plants and reemit some of it back toward the interior and the rest toward the exterior. The primary effect of greenhouses is thus to reduce heat loss by radiation toward the exterior. When they are closed and no air is allowed to circulate between the interior and exterior they also have a second effect of suppressing heat and humidity losses due to convection. The atmosphere plays a role analogous to that of garden greenhouses: it intercepts part of the infrared radiation emitted by Earth and sends some of it back toward the ground; it also nullifies all convective heat transfer toward the exterior by imposing a zero overall heat balance. The atmospheric greenhouse effect is estimated to generate about a heat flux of about 150 W m^{-2} toward the ground. This heat comes from the heat radiated by Earth, partially absorbed by the atmosphere, and partially reemitted back toward the Earth (see insert I1.5).

Gulf Stream - The Gulf Stream is the section of the thermohaline circulation situated in the North Atlantic Ocean between Western Europe and North America. The current comes from the southwest and influences the climate of the countries of Western Europe, which are relatively temperate in comparison with that of the United States or Canada at the same latitudes.

Harmonic - In the spectral decomposition of a wave, the integer multiples of the basic mode are called the *harmonics*, and the fractional multiples are called the *sub-harmonics* (see *Mode*).

Heterosphere - The heterosphere is the very high atmosphere, situated above the homosphere at over 100 km from the ground. The various chemical species present in the heterosphere are not as well mixed as in the homosphere.

Holocene - The Holocene is the geological era in which we now live. It began after the last ice age, which ended about 16 000 years ago.

Homosphere - The homosphere is the lower part of the terrestrial atmosphere, where chemical species are rather well mixed at all altitudes so that their fractional content in the atmosphere is constant. The homosphere itself consists of three main layers: the troposphere near the ground, the stratosphere, and the mesosphere, which are separated by transitions called the *tropopause* and the *stratopause* (see fig. 1.1).

Hull - The exterior surface of a ship that is in contact with the water.

Hydraulic jump - A hydraulic jump is a sudden change in the depth and speed of the flow of water with a free surface. It occurs in the sudden transition between the torrential regime upstream, where the slope of the flow bed and of the free surface is rather large, and the fluvial regime downstream where these slopes are rather small. This transition is very rapid and singular. The opposite and much more progressive transition from the fluvial regime to the torrential regime is said to be *regular* and does not result in a jump. The jump is analogous to a shock-wave; like a shockwave, it is accompanied by a significant dissipation of energy.

Ice age - Earth has known several significantly cold periods, which led to ice ages. During the Quaternary Period, four ice ages occurred, lasting between 60000 and 100000 years each. These are the Günz Era, which ended 540000 years ago, the Mindel Era, which ended 450000 years ago, the Riss Era, which ended 180000 year ago, and the Würm Era, which ended 16000 years ago. The enormous glaciers that formed during the last ice age, which were up to several thousand meters thick, dug valleys in which today flow large rivers such as the Rhine and the Rhone. The ice ages were separated by remissions, such as the Holocene in which we live today.

Ideal gas - An ideal gas is a theoretical model that represents rather well the state of real gas, provided it is rather rarefied (see *Equation of state*). Thus, the air in the troposphere, which is subjected to a pressure less than or equal to 1013 hPa at sea level, follows rather well the equation of state for ideal gases. However, under greater pressure, air departs rather significantly from the ideal gas law; to the point that corrective terms must be added to the ideal gas law, transforming it into the VAN DER WAALS equation.

Index of refraction - The refraction of a wave is the deviation of its trajectory due to a change in its celerity. It occurs when the density of the medium in which the wave propagates suddenly varies, such as upon crossing the interface between two different fluids (e.g., air and water). Each medium is characterized by an index of refraction and the deviation between the incident and refracted rays of light depends on the two indices according to the law $n_1 \sin\theta_1 = n_2 \sin\theta_2$, where n_1 and n_2 are the indices of refraction of each medium, and θ_1 and θ_2 are the angles between the incident and refracted rays and the surface normals. In a nonhomogeneous medium such as the air of the homosphere, the index can vary continuously, leading to a progressive curvature in light rays. The index of refraction also varies as a function of the wavelength of the radiation, which leads to rainbows and the reddening of the evening and morning skies.

Inertial waves - Rotating flows with a very small Rossby number are dominated by a balance between the Coriolis force and the pressure gradient, so any difference in relative speed between two planes perpendicular to the axis of rotation is transported by the axial component of the pressure gradient. Thus, this difference tends to be suppressed by these waves, whose main effect is to impose a tight correlation between the various planes perpendicular to the axis of rotation and thus to structure the flow into columns.

Infrared radiation - In the spectrum of radiation emitted by an object, the band corresponding to visible light is between 0.38 and 0.78 μm. Infrared radiation corresponds to longer wavelengths, going up to the hundred-micron range. It is absorbed rather strongly by air, which allows it to be used as a heat source on the scale of meters.

Infrasound and ultrasound - Human ears can detect acoustic waves only in the range from 15 Hz to 20000 Hz, which can vary from one individual to the next. Frequencies below 15 Hz constitute infrasound and those greater than 20000 Hz constitute ultrasound.

Instability - In mechanics the issue of stability is always present. So-called *unstable equilibrium* occurs when any perturbation of the system moves it irrevocably away from its equilibrium position. In contrast, an equilibrium is stable when any perturbation is met by a restoring force that tends to return the system to its equilibrium position.

Intermittency - This term refers to phenomena that appear or disappear rapidly, forming what can be referred to as localized bubbles in time or in space. In established turbulence, the smallest vortices that occupy only a very small fraction of the space are intermittent.

Intrados and extrados - These words designate the two faces of a wing in flight. The intrados (extrados) is the bottom (top) face of the wing; an overpressure (underpressure) is exerted on it when the wing has a positive angle of attack. Together, the overpressure on the intrados and the underpressure on the extrados explain the lift that allows the wing to fly. We can also speak of intrados and extrados for helicopter blades or even wind-turbine blades.

Ionization - Ionization is the phenomenon whereby one or several electrons are either removed from or added to an atom or molecule, leaving the atom or molecule with a net negative or positive charge, respectively.

Jet stream - This expression refers to an air current at high altitude (about 10 km above the ground) that moves at high speeds (about 300 km h^{-1}) from west to east, circumnavigating the globe at latitudes slightly above the tropics. The jet stream is present in both hemispheres and constitutes a sort of return current compensating for the westward-blowing trade winds that operate at lower altitudes near the equator. The ensemble of these currents is subjected to seasonal variations. The jet stream is subjected to rather strong instabilities that make it oscillate around an average trajectory.

Katabatic wind - In mountainous regions, this wind blows down the slopes during the morning, typically following the steepest combes. It is pumped by the lightening of the air in the valleys, which are less cooled during the night than the air in the high meadows and plateaus.

Kinetic theory of gases - This theory, which describes interactions between molecules of gas, derives the macroscopic properties (pressure, temperature, and all other variables that characterize the gas) based on a set of microscopic parameters (shape and motion of the molecules).

Latent heat - The heat or enthalpy that is received or furnished when matter at a given uniform temperature changes state (or *phase*). This quantity differs from the sensible heat, which is exchanged by conduction within matter at a nonuniform temperature. For water evaporating into air, the latent heat is withdrawn from the liquid near the liquid-vapor interface and then transported by conduction and convection. Thus, for boiling water at 100 °C and under normal pressure (1013 hPa), the latent heat furnished by external sources must be 2257 kJ kg^{-1}. Under the same conditions, when water vapor condenses, the same quantity of latent heat is liberated by water toward the exterior.

MACH number - This number is the ratio between the speed V of a fluid and the celerity c of acoustic waves in the fluid: $Ma = V/c$. For $Ma > 1$ ($Ma < 1$) the regime is supersonic (subsonic).

Magnetosphere - The magnetosphere is the region encircling a planet such as Earth, or a star, in which the lines of flux of its magnetic field loop back to form closed loops. Flux lines are everywhere tangent to Earth's magnetic field; they exit from the North Magnetic Pole (actually located near the Geographic South Pole of Earth), trace out a long path in the magnetosphere, and reenter the Earth at the South Magnetic Pole to connect up inside Earth. Without the solar wind, the terrestrial magnetosphere would form an essentially perfect axisymmetric dipole field. However, the speed of the solar wind blows the lines of flux away from the Sun, reducing the day side of the magnetosphere by about 5000 km and significantly lengthening the night side in the wake of Earth.

Mean free path - Within a gas, all molecules travel a certain distance between successive collisions with other molecules; this is called the *free path*. The term *mean free path* means the average of all the actual free paths. In air at normal pressure, the order of magnitude of the mean free path is about 100 nm.

Mesosphere - The mesosphere is the atmospheric layer that lies just above the stratosphere, between the about 55 and 100 km in altitude. Whereas the temperature increases with altitude in the stratosphere because of absorption of ultraviolet radiation, it falls strongly in the mesosphere to near −100 °C. At the surface of Earth, the mesosphere is noticeable due to the shooting stars that are produced when meteorites falling toward Earth disintegrate as they traverse it.

Mode - Vibrational phenomena, such as all types of waves, are often broken down into elementary modes, each of which correspond to a certain frequency or a certain wave number. The energy spectrum of a monochromatic vibration reveals frequency bands, among which we often find the principal mode of frequency f, integer multiples at frequencies $2f$, $3f$, $4f$... called *harmonics*, and fractional multiples at frequencies $f/2$, $f/3$, $f/4$... called *subharmonics*.

Monsoon - Summer monsoons occur when thermal winds, loaded with moisture after passing over a warm sea, arrive over a colder continent. The subsequent condensation can provoke torrential downpours. In contrast, winter monsoons occur when the ground cools off and the dried air is sucked away by the warmer air over the nearby ocean. If this winter wind can load up again with moisture by passing above a warm sea, it can then lead to rains. The monsoons are thus meteorological phenomena with a strong seasonal character.

Nadir - For a given observer, the nadir is the point opposite the zenith, exactly like south is opposite of north, or west vs east. The nadir is important for airline passengers or astronauts, who can look in its direction.

Order of magnitude - This expression is much used not only in physics but also in everyday language. It signifies an estimate rounded to the nearest factor of ten that is used instead of a precise value. Orders of magnitude allows for comparisons to be made between classes of phenomena of similar nature but of different scale.

Pairing - The association of two objects with similar characteristics that results in the formation of a pair and its subsequent development as a single entity. In large-scale two-dimensional flows such as occur in the atmosphere, low-pressure zones spanning several hundred kilometers form cyclonic eddies that turn in the direction of Earth's rotation and that pair up to form even-larger depressions. This process iterates until the friction limits its further evolution.

Penstock - In hydroelectric facilities with high hydraulic heads, the penstock channels water from the surge tank situated at approximately the same altitude as the free surface of the reservoir to the much lower altitude of the turbines. In the static state, the interior pressure varies strongly between the upper and lower extremities of this duct, and it can fluctuate significantly during variations in the flow rate. Penstocks are constructed in steel according to strict regulations and are solidly anchored in rock on steep slopes (see fig. 8.9).

Photon - Electromagnetic radiation has a double nature: it may be described both as a wave and as a flux of photons, which are particles of very small mass and energy (of the order of 2 eV for visible light). In our environment, all source of radiation, such as antennas and lamps, emit very large quantities of photons.

Rarefied gas - When the density of a gas is sufficiently low for the mean free path of molecules to become of the same order of magnitude as the characteristic length scales of flow (dimension of vortices, length of an airplane, diameter of a weather balloon…), the macroscopic model based on the classical equations of fluid mechanics, notably the NAVIER–STOKES equation, is no longer justified. The transport properties of such a gas must instead be studied within an appropriate kinetic theory, which is more complicated than the elementary model of insert I1.3.

RAYLEIGH number - The RAYLEIGH number is the essential parameter for determining the conditions under which a fluid heated from below remains stable. It is expressed as $Ra = (g\beta\Delta Th^3)/(v\alpha)$, where g is acceleration due to gravity, β is the thermal expansion coefficient of the fluid, v is the kinematic viscosity, α is the thermal diffusivity, ΔT is the temperature difference between the upper and lower interfaces of the fluid, and h is the distance between the interfaces.

Relative humidity - This is the ratio between the moisture content in air and the maximum possible moisture content, which is the saturation vapor pressure. A small relative humidity means that the air is far from being saturated with moisture and can still absorb more water vapor. A relative humidity of 100% means on the contrary that the air is saturated with moisture and that any further evaporation into this air is impossible.

Reference frame - In mechanics, two quantities are required to describe motion, one measuring the space in which the position of the moving object is measured, and the other measuring the time at which measurements are made. Together, these quantities form a reference frame. Among all the possible reference frames, *inertial* or *Galilean* reference frames are those in which NEWTON's second law holds. All other reference frames, and in particular that of the Earth as it moves through the Universe or that of a vehicle moving with respect to Earth, must be defined by its own motion with respect to an inertial frame. In such noninertial reference frames (i.e., accelerating reference frames), compensating terms must be added to NEWTON's second law, such as the CORIOLIS acceleration. To define the position of a point in any reference frame, a coordinate system must be chosen, such as Cartesian, spherical, cylindrical, or other.

REYNOLDS number - The REYNOLDS number is the ratio between inertial and viscous effects in a flowing fluid. It may be expressed as $Re = VL/v$, where V is the speed of the fluid, L is the characteristic length of the flow, and v is the kinematic viscosity of the fluid. When the REYNOLDS number is much greater than unity, viscous friction does not affect the equilibrium between forces except in very thin regions, such as in boundary layers. Outside these regions, only a pressure gradient and external forces (gravity, CORIOLIS force…) can balance the forces of inertia.

ROSSBY number - The ROSSBY number $Ro = V/\omega L$ is crucial for evaluating the pertinence of the CORIOLIS force for a rotating flow of typical length L and typical speed V that itself evolves in a rotating reference frame such as Earth.

This number is defined as the ratio of the period of revolution of the vortex ($\approx L/V$) and the characteristic timescale of the CORIOLIS force ($\approx 1/\omega$, where ω is the angular speed of rotation of Earth). For the large atmospheric depressions where the ROSSBY number can be less than 10^{-3}, the acceleration of the motion is negligible relative to the CORIOLIS acceleration. Thus, the regime is dominated by the geostrophic equilibrium (see definition).

Sensible heat - The heat exchanged within a mass of matter, or between two bodies, with no accompanying change of state (or *phase*). In opaque matter at rest, this heat is transported by conduction from the hotter zones toward the colder zones according to FOURIER's law. In a medium with internal motion the renewal of the fluid near the warm boundaries adds another contribution, and the combination of these two mechanisms constitutes the heat exchanged by conduction and convection. If the material is transparent, a third contribution is added by the radiance emitted toward all media visible from the emitting surface, as described by the STEFAN–BOLTZMANN law.

Shock wave - A shock wave is the superposition of acoustic waves that were emitted beforehand over a certain period of time. The faster waves overtake the slower ones and the energy of the ensemble becomes concentrated over a very thin space, forming what is called a *shock wave*. A shock wave is always accompanied by a jump in pressure, speed, and temperature between the two faces and by a dissipation of energy. A fraction of the kinetic energy dissipated by the shock wave is converted into acoustic energy, leading to the *sonic boom* of supersonic aircraft.

Similarity - For physical phenomena to be similar, their spatial boundaries must not only satisfy geometric similarity, but all actions that define them must also satisfy the conditions of similarity. In other words, for two instances of the same phenomenon to be similar, then the ratio of the values of any given parameter from the two instances must be the same for all parameters that define the phenomenon. In fluid mechanics, the theory of similarity is used very often to establish rules for designing scale models in the laboratory on which measurements can be made, the results of which will be representative of the actual, life-size phenomenon.

Solar radiation - The solar radiation arriving at the top of the homosphere covers a very large band of wavelengths going from about 0.1 to 10 μm. A significant part of its energy, essentially situated in the ultraviolet band, is absorbed in the stratosphere; the remaining fraction constitutes the source of energy responsible for the heating and motion of the principle terrestrial fluids, air and water, thereby distributing them over the planet.

Solar wind - This expression refers to the flux of ionized particles emitted in all directions by the Sun. Near this star, the solar magnetic field forces their trajectories to follow spiral paths. Near Earth and on its scale, this spiraling effect is no longer significant, but another magnetic field, this time from Earth's magnetosphere, serves to shield Earth from this flux of very energetic particles, preventing them from penetrating the atmosphere, except near the poles where this penetration leads to the polar auroras.

Spectrum - For all phenomena whose energy is spread over a certain wavelength range, the word *spectrum* refers to the contribution to the total energy from small intervals around each wavelength λ or wavenumber $k = 2\pi/\lambda$. The spectrum of such phenomena can thus be represented by a function $E(\lambda)$ or $E(k)$. We can speak of line spectra for phenomena whose energy is localized around one or several well-defined wavelengths, or continuous spectra for phenomena whose energy is distributed over rather large wavelength bands.

Sprites - Like the ELVES (see definition), these are luminous, extremely fleeting signatures of thunderstorms; they are sometimes called *sylphs*. They may be seen below ELVES at altitudes ranging from 50 km (the upper stratosphere) to 90 km (the lower mesosphere). They resemble shreds of cloth hanging and twisting from the stratosphere (see fig. 3.9).

Standard atmosphere - An atmosphere defined in the conventional manner by the International Civil Aviation Authority (ICAA) for average conditions very close to the average conditions in the real atmosphere, which is subjected to significant fluctuations. Aeronautical records are recorded in this standard atmosphere.

State equation - All quantities that characterize matter at rest macroscopically and that define its state are related by so-called *equations of state*. In a pure and sufficiently dilute gas, all quantities such as pressure p, temperature T, or density ρ are all three determined by fixing only two of these variables (which we choose to be the independent variables). These quantities are thus connected by a relationship that, in the case of air, is very close to the ideal-gas equation $p/\rho = RT/M$ where R is the universal gas constant and M is the molar mass of the gas. In a gaseous mixture of n species, the concentration of each is also a state variable. In this case, we simply add $n-1$ other variables to the two preceding ones and replace M by the mean molar mass of the mixture. All physical properties, such as viscosity, conductivity, and heat capacity, also allow us to verify the equations of state.

Stratopause - The stratopause is the region between the top of the stratosphere (see definition) and the bottom of the mesosphere; in other words, between the altitudes of about 50 and 55 km. In this zone, phenomena that dominate in the stratosphere diminish (notably the absorption of ultraviolet radiation that generates the ozone layer), and phenomena characteristic of the mesosphere are not yet well established (notably the strong decrease in temperature).

Stratosphere - The stratosphere is the layer of the homosphere situated between the tropopause, toward 8 to 15 km from the ground, and the stratopause, toward 50 km from the ground. Its main property is the warming of the air caused by the absorption of ultraviolet radiation from the Sun, which leads to a uniform increase in temperature from −56 °C in the tropopause to 0 °C in the stratopause. The density also decreases uniformly with altitude in the stratosphere. This layer is thus not subject to the convective RAYLEIGH–BÉNARD instability.

STROUHAL number - In the flow downstream of a cylindrical obstacle, if the REYNOLDS number is sufficiently large, vortices are released periodically from one side and then the other of the obstacle. The release frequency is such that, for a very large range of REYNOLDS number, the STROUHAL number $St = fD/V$, where f is the frequency, D is the diameter of the circular cylinder, and V is the speed of the distant flow, has a constant value close to 0.2.

Surge tank - In hydroelectric facilities with large hydraulic heads, the surge tank is situated between the approach duct, which is quite long, has a large diameter, and a small slope, and the penstock, which has a small diameter and which brings the flow directly to the gates and the turbines (see fig. 8.9). The equilibration path allows the gates to be rapidly closed without having to stop the entire mass of the water moving through the approach duct.

Sylphs - See Sprites.

TAYLOR column - Within a fluid held in a rotating container, all obstacles to flow tend to create a longitudinal perturbed domain in the fluid because of the effect of inertial waves, thereby forming a cylinder of rotating fluid aligned parallel to the axis of rotation of the container with its base fixed on the obstacle. When the ROSSBY number $Ro = V/\omega L$ is much less than unity, the inertial waves travel the distance L much faster than the fluid particles, thus managing to form a cylindrical column along which any variations in speed become negligible. Such columns can be observed in the atmosphere and in the oceans, where the scales are sufficiently large (see an example in fig A.9 in the appendix).

Thermal conductivity - All matter at rest macroscopically can nevertheless transport heat via the agitations of its atoms or molecules. This property of heat transport is called *conduction* and is characterized by a state variable κ called the *thermal conductivity*. Often, the thermal conductivity is replaced by the thermal diffusivity $\alpha = \kappa/\rho C_p$, where ρ is the density and C_p is the specific heat capacity at constant pressure.

Thermal equilibrium of the planet - When the heat received by the planet balances the heat that leaves the planet, a thermal equilibrium is reached and the average temperature becomes constant. However, any thermal imbalance leads either to an elevation or to a reduction of the temperature. Thus, in times of climate warming, the average temperature of Earth increases so that the infrared radiation emitted toward space exactly compensates the solar radiation that it receives.

Thermal wind - On Earth, all winds are directly or indirectly caused by sunshine and the associated temperature differences. They could thus all be called *thermal winds*. However, in practice, this term has come to mean the winds generated by sunshine hitting a rocky cliff or other significant feature of the landscape. The air near these steep slopes is warmed and thus lightened; it therefore rises and sucks in the surrounding air behind it. Thermal winds create local depressions, notably near mountain summits, and often result in the formation of small white cumulus clouds over summits. Paragliders and glider pilots search out the ascending air-flows created by thermal winds.

Thermocline - The thermocline refers to the transition between, on the one hand, the layers of water close to the free surface of a sea or lake, which are warmed by sunshine from spring to autumn and sufficiently mixed to an almost-uniform temperature, and, on the other hand, the deep layers that remain at a uniform temperature all year around. The thermocline builds up progressively in the spring, stabilizes in summer, progressively degrades in fall, and finally disappears in winter (see the Mediterranean thermocline in figure 5.1).

Thermodynamic equilibrium - When a mass of matter is at rest and exchanges neither mass, nor momentum, nor energy with the surrounding medium, it is said to be in *thermodynamic equilibrium*. It is precisely in this state of thermody-namic equilibrium that the equations of state hold, such as that for ideal gases. However, fluid particles that exchange momentum or energy only very slowly with the world external to the mass are typically assumed to be in a state of local equilibrium. This approximation is well justified when the characteristic timescale of the exchanges with the exterior (often of the order of seconds) is much larger than the characteristic timescale of molecular interactions (of the order of 10^{-12} s in a gas such as air).

Thermohaline circulation - The very large closed circuit of water flowing through the five oceans that takes over 1600 years to complete its round trip around the globe. The adjective *thermohaline* means that it is influenced by variations in density due to differences in temperature (thermo) and salinity (haline).

Tide - Tides are large displacements of fluid masses generated by the gravitational attraction of heavenly bodies. On Earth this phenomenon essentially concerns the ocean waters, which are attracted mostly by the Moon and the Sun. But the terrestrial crust, which has a certain plasticity, is also subject to this attraction, although the end effect is much weaker.

Trade winds - These are the winds from the tropical regions and result from the sunshine that is always more intense in regions where the Sun is at the zenith than in surrounding regions. The sunshine strongly heats an area on the ground of about 5000 km in diameter that, to follow the Sun, circumnavigates Earth along the equatorial and tropical latitudes. This part of the surface of Earth heats in turn the air above it, which rises and thereby sucks in the surrounding air, gen-erating a system of relatively constant winds called the *trade winds*.

Triboelectricity - The term refers to the physical phenomenon whereby rubbing a relatively good electrical insulator, such as glass or plastic, can lead to electrons being stripped from another material and fixed onto the new host.

Tropopause - This is the region that separates the troposphere from the stratosphere. The convective turbulence typical of the troposphere, which is heated from below by the ground, disappears in the tropopause and above because the stratosphere is heated from above.

Troposphere - This is the atmospheric layer closest to the ground, where the temperature decreases linearly as a function of altitude at an average of 6.5 °C km^{-1} and where the pressure decreases exponentially. Its thickness is of the order of 7 to 8 km at the poles and of the order of 15 km at the equator where the centrifugal force spreads it farther from the ground. The troposphere alone contains 80% of the total mass of the atmosphere.

Turbulence - This word refers to the random part of a fluid's motion. Turbulence thus represents the ensemble of fluctuations in speed, temperature, pressure, and all other state variables that, in a real flow, are superposed over the average values of the quantities.

Ultrasound - See Infrasound.

Ultraviolet radiation - Ultraviolet radiation corresponds to wavelengths below 0.38 µm. It is almost not absorbed by air, except in the stratosphere where it causes oxygen molecules to dissociate and form ozone.

Vapor pressure - This expression refers to the partial pressure of water vapor in air, which itself is the pressure that would be exerted in the given volume if it contained only the water vapor, with all other species of gas being removed. The term *saturation vapor pressure* refers to the upper limit of the vapor pressure above which air in thermodynamic equilibrium cannot accept more water in the gas phase.

Viscosity - Viscosity is a state variable of fluids and determines the fluid's capacity to create friction, which acts on all solid objects in displacement in the fluid, on the walls limiting the fluid domain and motion, and also within the fluid itself (in the boundary layers if the REYNOLDS number is sufficiently large). The work done by this friction systematically drains energy, which it transforms into heat that is evacuated by conduction or convection. Two varieties of viscosity exist: dynamic viscosity, which is the ratio between the tangential stress near a shear plane and the local speed gradient, and the kinematic viscosity, which is the dynamic viscosity divided by the fluid density. At 20 °C, the dynamic viscosity of water is about 10^{-3} Pa s, and that of air is about 1.8×10^{-5} Pa s.

Vortex intensity - The vortex vector is defined as $\boldsymbol{\omega} = (\boldsymbol{\nabla} \times \mathbf{V})/2$ so that its magnitude equals the angular speed of the local fluid particles. Such a vector is assigned to each point to describe a fluid in motion. Starting with the velocity, we can derive the kinetic-energy density of the fluid: $\rho V^2/2$. In a completely analogous fashion, the vortex vector $\boldsymbol{\omega}$ is associated with a scalar quantity $\omega^2/2$ called the *vortex intensity*, or *vorticity*.

Vorticity - See Vortex intensity.

Warm front - While circulating in the troposphere, masses of warm air may rub shoulders with masses of colder and thus heavier air. A warm front is the region where the warm, moist air is forced to pass above the cold and relatively stagnant air. Situated along the region at the interface between warm and cold air, a warm front is generally of rather small size. As the warm air gains altitude, it cools and its moisture condenses to form clouds, which can lead to precipitation or even to thunderstorms if the uplifting is sufficient.

Water hammer - Upon the sudden closure of a hydraulic gate or valve, acoustic waves propagate at high speed (about 1500 m s^{-1}) in the water and reflect from obstacles such as the duct walls near bends. The reflected waves may converge and strike the gate or valve like a sort of ram, creating a local pressure spike sufficiently strong to damage the equipment. The water-hammer phenomenon generates a characteristic noise.

Waterplane area - This is the area of water in a river or ocean that is outlined by the hull of a ship. Thus, the submerged volume of a ship is bounded by a closed area that includes the waterplane area.

Waterspout - This word refers to a relatively significant mass of water that is ejected from the surface of a sea, lake, or other waterway, either due to a tornado, or at the foot of a large waterfall.

Wave number - The wave number k is related to the wavelength λ in the same way as the angular speed ω is related to the period T: $k = 2\pi/\lambda$ vs $\omega = 2\pi/T$. These two quantities are used to construct the dimensionless arguments kx and ωt for wave functions.

Wavelength - The wavelength designates the spatial period of a wave, such as the distance between two successive crests of a swell or two successive troughs.

Index

© Springer International Publishing AG 2017
R. Moreau, *Air and Water*,
DOI 10.1007/978-3-319-65215-3